Contemporary Trends and Issues in Science Education

Volume 46

Series Editor
Dana L. Zeidler, *University of South Florida, Tampa, Florida, USA*

Founding Editor
Kenneth Tobin, *City University of New York, New York, USA*

SCOPE

The book series Contemporary Trends and Issues in Science Education provides a forum for innovative trends and issues connected to science education. Scholarship that focuses on advancing new visions, understanding, and is at the forefront of the field is found in this series. Accordingly, authoritative works based on empirical research and writings from disciplines external to science education, including historical, philosophical, psychological and sociological traditions, are represented here.

More information about this series at http://www.springer.com/series/6512

Mansoor Niaz

Evolving Nature of Objectivity in the History of Science and its Implications for Science Education

 Springer

Mansoor Niaz
Epistemology of Science Group,
   Department of Chemistry
Universidad de Oriente
Cumaná, Sucre
Venezuela

ISSN 1878-0482   ISSN 1878-0784   (electronic)
Contemporary Trends and Issues in Science Education
ISBN 978-3-319-88476-9   ISBN 978-3-319-67726-2 (eBook)
https://doi.org/10.1007/978-3-319-67726-2

Printed on acid-free paper

This Springer imprint is published by Springer Nature
The registered company is Springer International Publishing AG
The registered company address is: Gewerbestrasse 11, 6330 Cham, Switzerland

For Magda and Sabuhi

Preface

Like most science students I too was trained to understand that objectivity, certainty, truth, universality, and the scientific method were the five fundamental characteristics of both science and scientific progress. Despite all the reform efforts, science curricula and textbooks in most parts of the world continue to present science as a Baconian orgy of quantification. This inexorably leads students to believe that scientific progress depends on logically sound conclusions based on non-controversial experimental procedures. Considering that Kuhn's *The Structure of Scientific Revolutions* and Holton's *Introduction to Concepts and Theories in Physical Science* were published more than half a century ago, the present state of science education is all the more difficult to understand.

The relationship between objectivity, the scientific method, and inductivism had intrigued me for many years. About 6 years ago while teaching a course based on the role of history and philosophy of science (HPS) in teaching chemistry, one of the participating teachers expressed the following: "In contrast to the HPS perspective, the inductivist vision is rigid and does not contemplate *"transgressions* of objectivity."* In other words, besides empirical evidence we need to situate progress in science within a historical, cultural, and philosophical milieu of the time. Considering that the idea of "transgression" was not discussed in class, I found this to be a very novel idea. Similarly, about 4 years ago while teaching a course related to the role of creativity in science, one of the participants provided a very creative response to the question: Was Newton objective in the formulation of his theory? In order to respond this participant first alluded to how Newton's vision was molded by the work of Brahe, Copernicus, Kepler, and Galileo and then went on to state that thanks to Newton, Einstein could go beyond. This led the participant to formulate another question: was Einstein objective and responded: for how long? This approach struck me as that of approximating to an evolving nature of objectivity within a historical context. This book is dedicated to these two students (and others like them) who shared their thoughts with me and provided the incentive to keep exploring the difficult concept of objectivity.

Next, I was influenced by Ronald Giere's critique of those scientists and philosophers of science who consider that what drives scientists onwards is that there are *truths out there to be discovered*, and that such philosophical positions can be considered as "objectivist realism." Reading Lorraine Daston and Peter Galison's thesis of the evolving nature of objectivity within a historical perspective was a watershed event that provided me a sort of "eureka" experience. Their ideas eventually helped me to formulate the conceptual framework necessary for understanding objectivity in both science and science education. Roald Hoffmann's idea of "transgression of categorization" struck me as yet another way of approaching "transgression of objectivity" that in a sense facilitated a state of closure to my ideas on the subject. Furthermore, I found a common thread running through the ideas of Daston and Galison on the one hand and those of Hoffmann. Given the widespread use of objectivity–subjectivity dichotomy as almost a panacea, especially in science education, Glen Akenhead's suggestion that objectivity can be considered as an "opiate of the academic" seems plausible, and provided me with a new perspective.

It is important for me to have mentioned these experiences and how they helped me to understand objectivity and its evolving nature and thus provided the impetus for pursuing this subject for almost 10 years.

In writing this book, I did not have any particular course in mind. This has the advantage that the book could be adopted for various types of courses, such as teaching the nature of science, introduction to the history and philosophy of science, understanding the dynamics of scientific progress, and the evolving nature of objectivity. The intended audience for this book is secondary and university-level teachers, science teacher educators, researchers in science education, and graduate students.

Chapter 1 introduces the idea of "transgression of objectivity" and the evolving nature of objectivity within a historical perspective. A theoretical framework is presented in Chap. 2, based on Daston and Galison's (2007) ideas of truth-to-nature, mechanical objectivity, structural objectivity, and trained judgment. Understanding objectivity in research reported (1992–2014) in the journal *Science & Education* is the subject of Chap. 3. Next, Chap. 4 deals with understanding objectivity in research reported (1992–2015) in the *Journal of Research in Science Teaching*. Understanding objectivity in research reported in reference works related to science education is the subject of Chap. 5. The idea of science at a crossroads that is the relationship between transgression and objectivity in the context of nanotechnology is presented in Chap. 6. As a conclusion, Chap. 7 facilitates an understanding of the elusive nature of objectivity.

The following are some of the salient features of this book that can help readers to follow the line of argument developed in the different chapters:

1. A detailed account and evaluation (over a period of almost 25 years) of how the science education community conceptualizes objectivity.
2. Objectivity as a process and not a state, which can change/evolve continuously.
3. Objectivity and subjectivity can be considered as the two poles of a continuum.
4. The dualism between objectivity and subjectivity leads to a conflict in the evolving nature of objectivity.

5. Scientific facts are mute and hence need interpretation.
6. It is not the experimental data (Baconian orgy of quantification) but rather the diversity/plurality in a scientific discipline that contributes toward understanding objectivity.
7. Objectivity, certainty, truth, and infallibility as universal values of science may be challenged while studying controversies in their original historical context.
8. The scientific enterprise is characterized not by the scientific method, but rather controversies, alternative interpretations, ambiguity, uncertainty, and intuitiveness.
9. Open-mindedness and not relativity helps in understanding objectivity.
10. Reality presents a different perspective to different scientists and hence progress in science is based on narratives that generate tensions leading to "transgression of objectivity."
11. Polanyi's tacit knowledge represents trained judgment and logical positivism approximates to mechanical objectivity (based on the framework of Daston & Galison, 2007).
12. Scientists are probably less reflective of "tacit assumptions" that guide their reasoning than most other intellectuals of the modern age.
13. The tension between subjectivity and objectivity in assessment practices leads to the understanding that science involves interpretation (conceptual problems) and not just memorizing algorithms.
14. It is perhaps the contingent nature of science that manifests itself in the evolving nature of objectivity.
15. Scientific progress is at a crossroads due to the interaction between representation (passive measurement and observation) and presentation (intervention, active manipulation, nanotechnology).
16. Given the research reported in this book, science education is faced with the following dilemma: is objectivity an opiate of the academic?

Cumaná, Estado Sucre, Venezuela   Mansoor Niaz

Acknowledgments

Classroom experiences and interactions with my students provided the major source of inspiration for starting and later completing this book. My institution Universidad de Oriente (Venezuela) has provided support for research activities over the last many years. Peter Galison (Harvard University) was kind enough to read some of my preliminary ideas (Chap. 2) with respect to the role played by objectivity in various historical episodes and provided critical feedback. Roald Hoffmann (Cornell University, Nobel Laureate in chemistry) helped me to understand that his idea of "transgression of categorization" approximates to Daston and Galison's idea with respect to violating the rules of objectivity. Hoffmann read a preliminary and the final version of Chap. 6, and provided feedback that facilitated an understanding of the underlying relationship between transgression and objectivity. Glen Aikenhead (University of Saskatchewan) read the final version of Chap. 7 and provided insight by posing the question: is objectivity an opiate of the academic? Michael Weisberg (University of Pennsylvania) read the final version of Chap. 6 and suggested important changes. I am indebted to all these scholars for having responded to my queries and thus facilitated a better understanding of objectivity, its evolution in the history of science, and its implications for science education.

A special word of thanks is due to Dana Zeidler, Springer Series Editor; Bernadette Ohmer, Publishing Editor at Springer (Dordrecht); Claudia Acuna, Editor; and Marianna Pascale, Assistant Editor for their support, coordination, and encouragement throughout the various stages of publication.

Contents

Chapter 1
Introduction: Understanding Objectivity within a Historical Perspective

School and college science generally emphasize the inductive nature of science. This eventually leads to an image of objectivity that does not concur with the history of science. Similarly, the traditional perspective of science is that scientists ideally undertake their investigations without bias, prior beliefs, and presuppositions. According to Cawthron and Rowell (1978, p. 33), this image is based upon a Baconian conception of scientific method as a well-defined, quasi-mechanical process consisting of a number of characteristic stages: (a) observation and experiment; (b) inductive generalization; (c) hypothesis (the formulation of general scientific statements or laws); (d) attempted verification; (e) proof or disproof; and (f) objective knowledge. Based on these stages: "Each successful verification adds to the stock-pile of objective knowledge. And objectivity is ensured by the conceptual neutrality of the scientific statements, being based on observational and experimental evidence—on facts—presumed free from unfounded speculation or the constraints of tradition" (Cawthron & Rowell, 1978, p. 33). These authors consider this image of the scientist as inductivist-empiricist as a fantasy that is widely disseminated among students. Furthermore, scientific facts are mute and hence the need for interpretation. Medawar (1969) considers such presentations of science as a theatrical illusion and a travesty. Furthermore, Smolicz and Nunan (1975) suggested that science curricula are a "pernicious transfiguration" of what scientists actually do. At this stage it would interesting to contrast the normative ethics of science that should ideally guide scientific conduct as presented by Resnik (2010), a bioethicist:

> *Scientists should strive for objectivity in research and publication, and their interactions with peers, research sponsors, oversight agencies, and the public.* If one assumes that truth and knowledge are objective, then this norm also helps to promote science's epistemic goals of truthfulness and error avoidance. Strategies and methods designed to minimize bias and errors in research, such as good record-keeping practices, the peer review system, replication of results, and conflict of interest rules, are based on a commitment to objectivity. Scientists also have an obligation to strive for objectivity when giving expert testimony in court, or when serving on government panels and committees. (p. 153, italics in the original)

© Springer International Publishing AG 2018   1
M. Niaz, *Evolving Nature of Objectivity in the History of Science and its Implications for Science Education*, Contemporary Trends and Issues in Science Education,
DOI 10.1007/978-3-319-67726-2_1

A science student unaware of the history of science may endorse such epistemic virtues related to objectivity without any reservations. Almost 50 years ago, Nagel (1961) a philosopher of science had espoused a similar scenario in the pursuit for objective historical inquiry:

> Moreover, even if the social climate in which historians work did have a decisive influence upon their investigations, the prospects for objectively based conclusions in historical research would not therefore be necessarily hopeless, for the pursuit of objective historical inquiry might very well be one of the ideals prized and fostered by a society and controlling a historian's researches. (p. 581)

These statements (Nagel, 1961; Resnik, 2010) stand in sharp contrast and provide a backdrop to what scientists profess (based on declared strategies and methods to strive for objectivity), and with actual scientific practice. Science students, teachers, and textbook authors may also find such statements as an appropriate milieu for understanding the scientific enterprise. One of the objectives of this book is to precisely explore the degree to which such epistemic goals are practiced in the real world of science. Next, I contrast these views with those of scientists and historians like Darwin, Gould, Holton, Kuhn, Daston, and Galison.

As early as 1861, Charles Darwin had questioned the Baconian accumulation of data devoid of all theoretical considerations. Darwin critiqued the scientific procedure of earlier geologists in the following terms:

> About thirty years ago there was much talk that geologists ought only to observe and not theorize; and I well remember someone saying that at this rate a man might as well go into a gravel-pit and count the pebbles and describe the colors. How odd it is that anyone should not see that all observation must be for or against some view if it is to be of any service! (Letter written to Henry Fawcett, September 18, 1861, in Charles Darwin, Collected correspondence, 21 volumes. Cambridge University Press, Vol. 9, p. 269)

On reading this, Stephen J. Gould (1995) a paleontologist commented: "[Darwin] outlined his own conception of proper scientific procedure in the best one-liner ever penned. The last sentence is indelibly impressed on the portal to my psyche" (p. 148). Later Gould (1995) goes beyond by clarifying:

> Scientists often strive for special status by claiming a unique form of "objectivity" inherent in a supposedly universal procedure called *the* scientific method. We attain this objectivity by clearing the mind of all preconception and then simply seeing, in a pure and unfettered way, what nature presents. This image may be beguiling, but the claim is chimerical, and ultimately haughty and divisive. For the myth of pure perception raises scientists to a pinnacle above all other struggling intellectuals, who must remain mired in constraints of culture and psyche …. Objectivity is not an unobtainable emptying of mind, but a willingness to abandon a set of preferences—for *or against* some view, as Darwin said—when the world seems to work in a contrary way. (pp. 148–149, italics in the original)

First let us consider Darwin's views. Most science curricula and textbooks inculcate a view of science that comes quite close to "count the pebbles and describe the colors" and then leave the rest to the scientific method. Millions of students around the world study Darwin's theory of evolution and this raises the question: how many of these students are aware that Darwin also offered the following criticism: "all observation must be for or against some view if it is to be of any service." This also provides a good example of how domain-specific knowledge of the nature of science (NOS) that is evolutionary and geological theories can be integrated with domain-general NOS, namely theory-laden nature of observations (for details with respect to the integration of the two NOS aspects, see Niaz, 2016). Next, let us consider Gould's views. Scientists' claim to objectivity is based on a universal procedure based on the scientific method. This enables the scientists to observe nature without preconceptions and constraints. Gould considers such views not only divisive but also haughty. On the contrary, he suggests that objectivity is the "willingness to abandon a set of preferences" or in Holton's (1978a, b) perspective "willingness to suspend judgment."

Thomas Kuhn has generally endorsed the traditional view for evaluating the adequacy of a scientific theory based on: accuracy, consistency, scope, simplicity, and fruitfulness (Kuhn, 1977, pp. 321–323). Actually, Kuhn's views on objectivity and subjectivity are much more complex and he is often considered as responsible for depriving science of objectivity: "My point is, then, that every individual choice between competing theories depends on a mixture of objective and subjective factors, or of shared and individual criteria. Since the latter [subjectivity] have not ordinarily figured in the philosophy of science, my emphasis upon them has made my belief in the former hard for my critics to see" (Kuhn, 1977, p. 325). Kuhn specifically refers to the difficulties involved in applying the traditional criteria for theory choice, for instance in the following historical episodes: Ptolemy's and Copernicus's astronomical theories, oxygen and phlogiston theories of combustion, Newtonian mechanics, and the quantum theory. It is plausible to suggest that the evolving nature of objectivity (see later section in this chapter), as developed by Daston and Galison (2007), attempts to redress this lack of attention to the role of subjectivity in the philosophy of science literature. Finally, Kuhn (1977) concluded his arguments in the following terms: "It first provided evidence that the choices scientists make between competing theories depend not only on shared criteria—those my critics call objective—but also on idiosyncratic factors dependent on individual biography and personality. The latter are, in my critics' vocabulary, subjective …. What the tradition sees as eliminable imperfections in its rules of choice I take to be in part responses to the essential nature of science" (pp. 329–330). This clearly shows the importance of situating the objectivity-subjectivity debate within the context of understanding nature of science, which forms an important part of current research in science education (cf. Niaz, 2016; Chap. 3).

Although objectivity is not synonymous with truth or certainty, it has eclipsed other epistemic virtues, and to be objective is often used as a synonym for scientific in both science and science education. According to Daston and Galison (2007), the history of objectivity is nothing less than the history of science itself and the evolving and varying forms of objectivity does not mean that one replaced the other in a sequence but rather each form supplements and not supplants the others (p. 318).

Research in science education has emphasized the importance of nature of science (NOS) as a series of domain-general and domain-specific aspects based on historical scrutiny of the scientific endeavor, recognized in various parts of the world (AAAS, 1993; Abd-El-Khalick, 2012; Chang, Chang, Chang, & Tseng, 2010; Deng, Chai, Tsai, & Lin, 2014; Hodson & Wong, 2014; Lederman, 2007; Lederman, Abd-El-Khalick, Bell, & Schwartz, 2002; McComas, Clough, & Almazroa, 1998; Niaz, 2009, 2016; NRC, 2013; Smith & Scharmann, 2008; Vesterinen & Aksela, 2013). For example, a domain-general aspect of science would be the tentative nature of scientific knowledge and the changing nature of atomic models would represent the domain-specific aspect of NOS. There is, however, considerable controversy with respect to these descriptors, both in the science education and philosophy of science literature:

> While most scientists would likely agree that these descriptors accurately characterize their work, in recent years philosophers of science have recognized these criteria as simplistic and grossly inadequate for distinguishing between science and nonscience. The argument can clearly be made, for example, that scientists are human beings and that both the questions they ask and the interpretations they place on their data are influenced, albeit usually unconsciously, by their own personal histories and the prevailing disciplinary paradigms (Kuhn, 1970). Therefore, *science cannot be unequivocally objective.* (Smith, Siegel, & McInerney, 1995, p. 29, italics added)

Interestingly, a recent study has highlighted the need for science teachers to go beyond the myth that "seeing is believing," namely the objective nature of science in cogent terms:

> It is still not common for teachers to discuss the ways in which experiments, as well as observations, are theory impregnated or to point out that we can only investigate what we have speculated about, and in terms of *how* we have speculated about them. In a sense, as our respondents [practicing scientists] repeatedly told us, theoretical assumptions bias the inquiry and prejudice the conclusions. In consequence the notion of absolute scientific objectivity is a myth. Observational and experimental data do not "speak for themselves"; all data have to be interpreted. (Wong & Hodson, 2009, p. 124, italics in original, underline added)

It is important to note that this study is based on 13 well-established and practicing scientists from different parts of the world, in a wide range of specialized fields such as astrophysics, experimental particle physics, molecular biology, and cancer research. In a similar vein, Schwab (1974) has emphasized the role played by "heuristic principles" both in understanding and teaching science:

> A fresh line of scientific research has its origins not in objective facts alone, but in a conception, a deliberate construction of the mind. On this conception, all else depends. It [heuristic principle] tells us what facts to look for in the research. It tells us what meaning to assign these facts. (p. 164)

As suggested by Holton (1969), science textbooks, curricula, and classroom practice do just the opposite by emphasizing "experimenticism," namely progress in science is presented as the inexorable result of the pursuit of logically sound conclusions from unambiguous experimental data.

In both science and science education there is also a general perception with respect to the illusion of objectivity in statistical analysis. Berger and Berry (1988), however, have argued that although objective data can be obtained, reaching sensible conclusions from statistical analysis of data may require subjective input:

> This conclusion is in no way harmful or demeaning to statistical analysis. Far from it; to acknowledge the subjectivity inherent in the interpretation of data is to recognize the central role of statistical analysis as a formal mechanism by which new evidence can be integrated with existing knowledge. Such a view of statistics as a dynamic discipline is far from the common perception of a rather dry, automatic technology for processing data. (p. 159)

This facilitates an understanding of the interaction between objectivity and subjectivity and forms an important part of the Daston and Galison's (2007) framework, which is the subject of Chap. 2.

1.1   Transgression of Objectivity

This section draws on a study based on 26 in-service chemistry teachers who had enrolled in the course, "Investigation in the Teaching of Chemistry" as part of a Master's degree program in education at a major university in Latin America (for complete details, see Niaz, 2012; Chap. 5, pp. 149–178). Ten teachers worked in secondary schools and 16 at the university level (female = 16, male = 10, age range: 25–40 years), and their teaching experience varied from 5 to 15 years. In the previous year all teachers had enrolled in the course, "Methodology of Investigation in Education," in which the basic philosophical ideas of Popper, Kuhn, Lakatos, and Giere were discussed in order to provide an overview of the controversial nature of progress in science and its implications for research methodology in education. Teachers were aware that basic ideas like the scientific method, objectivity, and the empirical nature of science were considered to be questionable by philosophers of science. The course was based on 18 required readings that dealt among others with the following topics: (a) History and philosophy of science in the context of the development of chemistry; (b) Students' alternative conceptions; and (c) Conceptual change in learning chemistry. Class activities were based on discussions, oral presentations by the teachers, written exams and a take-home critical essay. Among other subjects the following aspects related to the nature of science (NOS) were discussed (similar to that included in the literature cited in the previous section):

1. Scientific theories are tentative.
2. Theories do not become laws even with additional empirical evidence.
3. All observations are theory-laden.
4. Science is objective only in a certain context of scientific development.
5. Objectivity in science is based on a social process of competitive cross-validation through critical peer review (Campbell, 1988a, b).
6. Science is not characterized by its objectivity but, rather, by its progressive character—progressive problemshifts (Lakatos, 1970).

7. Progress in science is characterized by conflicts, competitions, inconsistencies, and controversies among rival theories.
8. Scientists can interpret the same experimental data in more than one way.
9. Most of the scientific laws are irrelevant and at best can be considered as idea-
   lizations (Giere, 1999, 2006b).
10. There is no universal scientific method based on steps to be followed.

With this experience the teachers were asked to respond to the following question: "Given the importance of the different aspects of the nature of science, in your
opinion: what are the factors that impede the implementation of new strategies in
teaching chemistry?" (Reproduced in Niaz, 2012, p. 163). Teachers responded by
referring to the following factors (teachers could provide more than one factor): (a)
Empiricist presentations in chemistry textbooks that lack a history and philosophy
of science (HPS) perspective ($n = 19$); (b) Schwab's "rhetoric of conclusions" and
lack of "heuristic principles" ($n = 18$); (c) Teachers' and students' epistemological
views ($n = 16$); (d) Scientific progress devoid of controversy, interpretation of data,
and idealization ($n = 14$); (e) Non-recognition of the tentative nature of scientific
knowledge ($n = 11$); (f) Curricular programs and reforms ($n = 8$); (g) Empirical and
not theory-laden nature of science ($n = 8$); (h) Objectivity in science based on
experimenticism ($n = 7$); and (i) Scientific method ($n = 6$). Many of these responses
overlap and could have been classified differently. What is important, however, is to
note that most of the teachers had a fairly good understanding of the scientific
endeavor itself and how it could be included in classroom discussions. Interestingly,
one of the teachers selected factor (a) and provided the following justification:

> In contrast to the HPS perspective, the inductivist vision is rigid and does not contemplate
> "*transgressions*" of objectivity, precision, and methodology. This rigidity makes any
> change in a paradigm difficult, and textbooks continue to repeat the same framework.
> (Reproduced in Niaz, 2012, p. 164, italics added)

This response has various novel features such as contrast between the inductivist and the HPS perspective, rigidity of paradigms (perhaps evoking Kuhn), role
of textbooks, and "*transgression*" of objectivity. Except for the idea of transgression all the other aspects in this response were discussed in an explicit or implicit
manner during classroom activities. A question that comes to mind is: how did
this teacher come up with this idea of associating "transgression with objectivity"?
Of course, I could not follow up as I read the written responses after some time
and the participating teachers had by then left for their respective institutions
(some were from neighboring cities). In part, the need and the intellectual curiosity
to understand this teacher's response led me to undertake the present study.
Understanding "transgression of objectivity" is important and leads to a dilemma:
If scientists are absolutely "objective" then the path from data to theory (or for
that matter vice versa) would be free from controversy. However, philosophers
of science have referred to this as a "paradoxical dissociation" and explained in
the following terms: "While nobody would deny that *science in the making* has
been replete with controversies, the same people often depict its essence or end
product as free from disputes, as the uncontroversial rational human endeavor par

excellence" (Machamer, Pera, & Baltas, 2000, p. 3, italics added). The reference to *science in the making* in the above quote is very helpful in understanding the role of objectivity in scientific progress. Similarly, science educators and textbooks present a vision of science (a false image) that is free of controversies. Let us recapitulate this line of reasoning: history of science is replete with controversies and still the scientists (among others) ignore them and hence it follows that scientists are objective. In a way scientists consider that by ignoring controversies they can show that they are objective. This suggests that one way of understanding objectivity is precisely a historical reconstruction of the different phases that constitute scientific progress. This insight was important in the development of arguments in this book that led to an understanding of objectivity.

More recently, I was intrigued further on reading about the idea of "transgression of categorization" in Roald Hoffmann (2012, p. 36), a Nobel Laureate in chemistry. The relationship between "transgression of objectivity" and "transgression of categorization" and Hoffmann's ideas will be elaborated and discussed in detail in Chap. 6.

1.2   Evolving Nature of Objectivity

One way of understanding objectivity is precisely a historical reconstruction of scientific progress in which controversies are highlighted. Furthermore, this historical perspective reveals the evolving nature objectivity. On the contrary, in both science and science education to be objective is often used as a synonym for scientific. Daston and Galison (2007) have constructed the evolving nature of this scientific judgment through the following phases: truth-to-nature (eighteenth century), mechanical objectivity (nineteenth century), structural objectivity (late nineteenth century), and finally trained judgment (twentieth century). In truth-to-nature, objects were depicted not as particulars but universals that are a form of idealization. Mechanical objectivity made a virtue of attending to particulars and a vice of idealization. In trained judgment, the observer was an expert based on long and careful training that helped to eliminate artifacts of the instruments and categorize the world. Each of these regimes did not supplant the other but they can coexist and even supplement each other at the same time. It is important to note that the essential aspects of the history of scientific objectivity were first presented by these authors in the following terms:

> As historians of objectivity, we will not be concerned with recent controversies over whether objectivity exists and, if so, which disciplines have it. We believe, however, that a history of scientific objectivity may clarify these debates by revealing both the *diversity and contingency* of the components that make up the current concept. Without knowing what we mean and why we mean it in asking such questions as "Is scientific knowledge objective?," it is hard to imagine what a sensible answer would look like. (Daston & Galison, 1992, p. 82, italics added)

Indeed, it is plausible to suggest that the diversity and contingency of how objectivity came to be associated with scientific knowledge is equally important for science education as well.

At this stage it would be helpful to contrast the ideas of Daston and Galison (2007) with those of most science textbooks and even some historians and philosophers of science:

> The very idea of the modern natural sciences is bound up with an appreciation that they are objective rather than subjective accounts .... The objective character of the natural sciences is supposed to be further secured by a method that disciplines practitioners to set aside their passions and interests in the making of scientific knowledge. Science, in this account, fails to report objectively on the world—it fails to *be* science—if it allows considerations of value, morality, or politics to intrude into the processes of making and validating knowledge. When science is being done, society is kept at bay. The broad form of this understanding of science was developed in the seventeenth century, and that is one major reason canonical accounts have identified the Scientific Revolution as the epoch that made the world modern. (Shapin, 1996, p. 162, original italics)

Shapin's critique clearly shows the need to distinguish how science needs to be practiced from how it is actually practiced. It is in the latter context that both scientists and science educators ignore the complexities involved in scientific progress. Despite the similarities, there are some differences in the accounts of Shapin (1996) and Daston and Galison (2007) with respect to when this understanding of science originated—that, however, is not the subject of this study.

In order to understand further how science is actually practiced and understood, it is interesting to consider the following statement from Ziman (2000):

> Contrary to the Legend, science is not a uniquely privileged way of understanding things, superior to all others. It is not based on firmer or deeper foundations than any other mode of human cognition. Scientific knowledge is not a universal "metanarrative" from which one might eventually expect to be able to deduce a reliable answer to every meaningful question about the world. It is not objective but reflexive: the interaction between the knower and what is to be known is an essential element of the knowledge. And like any other human product, it is not value-free, but permeated with social interests. (p. 327)

This approximates not to relativism but on the contrary leads to an understanding of scientific knowledge as fallible. Changes in Newtonian mechanics based on Einstein's theory of general and special relativity show how our idea of universal knowledge undergoes modifications. Furthermore, the interaction between the knower and knowledge recognizes the role of the scientific community.

In order to facilitate a better understanding of the issues involved, at this stage it would be helpful to present the philosophical perspective of the postpositivists (Johnson & Onwuegbuzie, 2004; Phillips & Burbules, 2000): (a) what appears reasonable can vary across persons; (b) theory-ladenness of observations; (c) same experimental data can be explained by different theories; (d) the Duhem-Quine thesis; (e) empirical evidence does not provide conclusive proof; and (f) attitudes, beliefs, and values of the researchers influence their findings, so that fully objective and value-free research is a myth.

Although all scientific observations involve the use of instruments, there is no such thing as a perfectly transparent instrument. All instruments are limited to recording a few aspects of the observations they study and that too with a limited accuracy. Scientific instruments do not reveal the universality of science and thus for example, "There is no such thing … as *the* way the Milky Way looks. There is only the way it looks to each instrument …. There just is no universal instrument that could record every aspect of any natural object or process" (Giere, 2006a, p. 30, original italics). Furthermore, Giere (2010) has argued that knowledge claims are perspectival rather than absolutely objective and hence cannot provide a "true" or "correct" answer to a problem. Based on this understanding it is plausible to suggest that this leads to a *pluralism of perspectives* (Giere, 2006b) that facilitate a better understanding of different and rival interpretations (diversity of knowledge claims) accepted by the scientific community. Thus, it follows that the strongest claims a scientist can legitimately make are of a qualified and conditional form. At this stage it would be interesting to contrast Giere's perspective with that of Agazzi (2014): "… one now finds another no less deeply rooted perspective— among professional scientists as well as various cultivated people—namely, the belief that the assertions of science, though not deserving simply to be called *true*, must nevertheless be considered *objective*" (p. 1, original italics). In other words, Agazzi is willing to forsake the truth of a theory but not its objectivity. Giere (2006b) shows that such a position is not tenable: "By claiming too much authority for science, objective realists misrepresent science as a rival source of absolute truths, thus inviting the charge that science is just another religion, another faith. A proper understanding of the nature of scientific investigation supports the rejection of all claims to absolute truths" (p. 16). Furthermore, before the success of relativity theory and quantum mechanics, many physicists believed that classical mechanics was objectively true.

Relationship between truth and objectivity of scientific theories and its problematic nature has been explored by Niaz (2016, Chap. 3). This study is based on 12 in-service science teachers who had enrolled in the following required course: "Science, technology, ethics, and creativity in research" as part of their doctoral degree program at a major university in Venezuela. All participants responded to the following question as part of their formal evaluation for the course:

> Many scientists, science textbook authors and professors believe that science is "objective." If we accept this perspective, Newton's laws constitute the best example of objectivity in science. Nevertheless, at the beginning of the 20th century, Einstein's theories of relativity (special and general) questioned Newton's laws. Accordingly, do you think that Newton's laws are false and consequently that he was not "objective"? (Reproduced in Niaz, 2016, p. 60)

Most philosophers of science (including Duhem, Giere, Kuhn, Lakatos, and Laudan) would agree that if a scientific theory is replaced by another with greater explanatory power, it does not mean that the previous theory was either false or that its author was not being "objective." This is the dilemma faced by the participants in this question. In other words, Newton's laws when first proposed in the seventeenth century were "true" for that time (actually for more than 200 years) and he was as "objective" as one could possibly expect a scientist to be. Consequently, the solution to the dilemma lies in recognizing that both Newton and Einstein were being "objective" and provided theories that varied in their explanatory power in certain domains (e.g., Einstein explained better the behavior of particles approaching the velocity of light). With this background it is easier to understand the responses provided by the participants of this study. It seems that a majority (10 out of 12) of the participants had a fairly good understanding of the role of "truth" of a theory and consequently the "objectivity" of the scientist. Following Giere (1999, 2006a, b); scientific theories are not necessarily "true" or "false" and similarly the role of the scientist is more perspectival rather than "objective." Although this may seem to be a difficult question, most participants took considerable interest in responding and following is one example:

> First it is important to recognize that Newton molded his vision of the material world based on the law of universal gravitation, thanks to the work of scientists such as T. Brahe, N. Copernicus, J. Kepler, and G. Galilei. Was Newton objective in the formulation of his theory? He thought that he was and many believed that his vision was the last word with respect to this problem. However, Einstein demonstrated with his theory of relativity that Newton was not sufficiently objective as his theory could not explain certain phenomena that the theory of relativity could. But thanks to Newton, Einstein could see beyond Newton. Are Newton's laws false? In physics it is known that these laws are not fulfilled in the context of Einstein's physics and consequently are not objective in this context. Nevertheless, these days Newton's laws continue to be applied, and consequently, I think that in a certain sense these laws have "some degree of truth" in their natural context of application. Was Einstein objective? Until now history tells us that he was. For how long? We still do not know (Participant #2, Reproduced in Niaz, 2016, p. 65).

Background to this item is provided by Giere's (2006a, b) critique of those scientists and philosophers of science who consider that what drives scientists onwards is that there are *truths out there to be discovered*, and that such philosophical positions can be considered as "objectivist realism." Participant #2 tried to understand Newton's contribution in an evolving historical context by recognizing the work of Brahe, Copernicus, Kepler, and Galileo, which is a sound approach. However, this participant was clearly struggling to understand the dilemma, as she/he asked, "Was Newton objective in the formulation of his theory?" and again responded in a historical context by pointing out that, "many believed that his vision was the last word with respect to this problem." Next this participant reminded us that "But thanks to Newton, Einstein could see beyond Newton," and this helped to respond to the question, "Are Newton's laws false?" Finally, this participant raised a thought-provoking question, "Was Einstein objective?" and responded laconically, "For how long?" In my opinion, this line of reasoning (especially the reference to: for how long) approximates to an evolving nature of objectivity very much within a historical context as suggested by Daston and Galison (2007). It is important to note that no mention of the Daston and Galison thesis was made during class discussions or in the suggested readings. This suggests that a historical reconstruction facilitates a perspective that is conducive toward an evolving nature of "objectivity and truth."

Following this line of argument, Holton (2014a) has gone one step further by pointing out that, "The squishy phrase 'understanding of science' can mean many things, but above all it must, I insist, include *knowledge* of science, plus an acquaintance with <u>how science is done</u>, plus a view of science as part of the cultural development of humanity" (p. 1876, italics in the original, underline added). *How science is done*, approximates to what Shapin (1996) had referred to as how science is actually practiced. Based on specific episodes in the history of science a recent study has endorsed the changing/evolving meanings of objectivity: "By contrast, historians of science have offered rich historical analysis that aim to clarify the changing historical meanings of objectivity by examining the emergence of particular scientific ideals in specific episodes in the history of science. These historical studies have revealed *the complex, multifaceted, and ultimately contingent nature of the ideals that contribute to our current notions and understandings of scientific objectivity*" (Tsou, Richardson, & Padovani, 2015, p. 2, italics added).

Again, *cultural development of humanity* can be understood differently. For example, Harding (2015) advocates a philosophy of science for all research disciplines which permits a form of objectivity allied with a deep concern for social justice. Furthermore, she contends that objectivity and certain forms of diversity can be mutually supportive and that objectivity is too powerful a concept to be abandoned.

Similarly, according to Machamer and Wolters (2004): "… to save the objectivity of science, we must free it from an ideal of rationality modeled after mathematics and logic; we must show that both rationality and objectivity come in degrees and that the task of good science is to increase these degrees as far as possible" (pp. 9–10). This coincides with the perspective of Daston and Galison (2007) with respect to the evolving nature of objectivity. Similar ideas with respect to objectivity are difficult to accept in science education. In a sense this book explores the present status of objectivity in science education and how it can develop further in order to deepen our understanding of the scientific endeavor.

At this stage it would be interesting to consider other accounts of scientific objectivity that contrast with that of Daston and Galison (2007) and following is an example:

> The natural task of our knowing is indeed that of "grasping" reality; and abstractly speaking, we should say that such a goal is reached with the obtaining of "objective knowledge" that is, knowledge which matches that portion of reality that is its purpose to match. But, on the other hand, man seems always to be afraid of not being able to complete such a task; and doubts regarding this matter come from the fact that very frequently different persons, confronted with the same portion of reality, describe it in different ways. The conclusion is easy: *if different pictures are proposed concerning the same reality*, none of them (or possibly just one) can be "objective," that is can "correspond to the object," whereas all of them (with one possible exception) must be considered as purely "subjective"—as expressing a certain way of envisaging objective reality which is typical of some single subject. (Agazzi, 2014, pp. 51–52, italics added)

The picture of "objective knowledge" presented by Agazzi is quite at odds with that of Daston and Galison (2007). Agazzi considers that objective knowledge is primarily achieved by grasping reality, and gives the impression that this is unproblematic. Interestingly, at the same time Agazzi considers the possibility of "*if different pictures are proposed concerning the same reality*" (the part in italics). However, history of science shows that very frequently scientists present different pictures of the same reality leading to controversies. According to Machamer *et al.* (2000), despite beliefs to the contrary, science in the making is replete with controversies. This clearly shows that grasping reality is problematic and requires considerable clarification before consensus is achieved. It is precisely for this reason that Daston and Galison (2007) consider the historical evolution of objectivity as the history of science itself. Let us now consider Agazzi's (2014) solution to the problem of *different pictures being proposed concerning the same reality*: only one picture is objective and the remaining are purely subjective. This leads to yet another conundrum: how do we decide which the objective picture is? It is precisely in this context that the intricate relationship between subjectivity and objectivity becomes important. It is plausible to suggest that the exploration of the subjectivity of an individual self facilitates objectivity. Consequently, subjectivity is not a weakness of the self to be corrected or controlled (Daston & Galison, 2007, p. 374). For example, in the history of science, trained judgment as a reaction to mechanical objectivity was precisely based on the recognition of the role played by subjectivity.

At this stage it is important to note that the evolving forms of objectivity in the history of science provide a deeper understanding of scientific progress. Based on this understanding, elaboration of criteria for evaluating research in science education is more meaningful. In this chapter I have compared and contrasted the views of various philosophers of science (e.g., Nagel, Agazzi versus Daston, Galison, and Giere) to show the complexity of the issues involved and how science educators face the complex task of understanding the evolving nature of objectivity in the history of science. Despite these difficulties, I have also provided examples from two studies (Niaz, 2012, 2016) to show that given the appropriate milieu, science educators can understand "transgression of objectivity" and the underlying issues.

In a recent study, Galison (2015a) has extended their understanding of the historical evolution of objectivity in science (Daston & Galison, 2007) to the field of journalism. It would be of interest to see how this history of objectivity is reflected in the field of science education, given its close ties with the history of science itself. Based on these considerations this book has the following objectives:

1. Explore the evolving forms of scientific judgment including objectivity in the history of science as suggested by Daston and Galison (2007).
2. Based on this exploration related to objectivity, elaborate criteria for evaluating research in science education, within a history and philosophy of science framework.
3. Based on these criteria, evaluate research published in the following sources: *Science & Education* (Springer journal) in the 23-year period (1992–2014), *Journal of Research in Science Teaching* (Wiley-Blackwell journal) in the 24-year period (1992–2015), *International Handbook of Research in History, Philosophy and Science Teaching* (2014, Springer), and *Encyclopedia of Science Education* (2015, Springer);
4. Evaluate general chemistry textbooks published in the USA, based on a series of five criteria related to objectivity, scientific method, transgression of objectivity, and nanotechnology (as suggested by Daston & Galison, 2007 and Hoffmann, 2012).

The rationale behind these four objectives is precisely the importance of understanding the evolving nature of objectivity in a historical context, which facilitates the elaboration of criteria for evaluating research in science education. Evaluation of general chemistry textbooks provides the opportunity for exploring educational implications of "transgression of objectivity" and the relationship between "representation and intervention" (Hacking, 1983) through nanotechnology. At this stage it would be interesting to consider further studies based on curriculum reform documents (e.g., ACARA, 2015; CMEC, 1997; NRC, 2013) from different countries, as suggested by one of the reviewers of this book.

1.3 Chapter Outlines

The objective of the chapter outlines is to provide the reader an overview of the different chapters by including some salient features. Some outlines are longer, due to the length of the chapter.

Introduction: Understanding Objectivity within a Historical Perspective (Chap. 1). The traditional conception of science and science education considers that objectivity of scientific statements is ensured as these are based on experimental facts. History of science, however, shows that this inductivist stance is at best a fantasy. Objectivity consists in the willingness to abandon a set of preferences when faced with contrary evidence. Although objectivity is not synonymous with truth or certainty it is often used as a synonym for scientific. History of objectivity is nothing less than the history of science itself. The notion of an absolute scientific objectivity is a myth. Any change in science textbooks or curricula is difficult as the inductivist vision is rigid and does not contemplate "transgressions" of objectivity. One way of understanding objectivity is precisely a historical reconstruction of scientific progress in which controversies are highlighted. This historical perspective reveals the evolving nature of objectivity. Daston and Galison (2007) constructed the evolving nature of scientific judgment (objectivity) through the following phases: truth-to-nature (eighteenth century), mechanical objectivity (nineteenth century), structural objectivity (late nineteenth century) and finally trained judgment (twentieth century). This reconstruction shows the need to distinguish how science needs to be practiced from how it is actually practiced. A major difficulty is based on recognizing that scientific instruments do not reveal the universality of science. Based on the instrument used the strongest claims a scientist can make are of a qualified and conditional form. Giere (2006a, b) has presented a critique of those scientists and philosophers of science who consider that what drives scientific research is the pursuit of truth and that such philosophical positions can be considered as "objectivist realism." For example, before the success of relativity theory and quantum mechanics, many physicists believed that Newtonian mechanics was objectively true. Holton (2014a) goes beyond by emphasizing a view of science as part of the cultural development of humanity. This book is based on the premise that a historical reconstruction facilitates a perspective that is conducive toward an evolving nature of objectivity. A major objective of this book is to explore the presentation of objectivity in different sources (journals, handbook, encyclopedia, and textbooks) of interest to science educators.

Objectivity in the Making (Chap. 2). The theoretical framework of studies reported in this book is based on an examination of the evolving forms of scientific judgment (including objectivity) in the history of science as suggested by Daston and Galison (1992, 2007). Scientists who followed truth-to-nature were looking for the idea in the observation and not the raw observation itself. For example, the procedures for describing, depicting, and classifying plants were openly selective. Later, mechanical objectivity considered such drawings as subjective distortions. Those following mechanical objectivity called for objective photographs to supplement, correct, or even replace the subjective drawings. The controversy between two histologists in the late nineteenth century, Santiago Ramón y Cajal from Spain and Camillo Golgi from Italy, is quite representative of the issues involved in mechanical objectivity and truth-to-nature, respectively. Cajal defended his undistorted sight and charged Golgi of having intervened deliberately in accordance with his theoretical presuppositions. Interestingly, both got the 1906 Nobel Prize for Medicine. In the early twentieth century, many scientists became convinced that subjectivity was difficult to separate from objectivity and some became skeptical of scientific photographs and instead started to look in the domain of mathematics and logic, namely structural objectivity. Structures could be communicated to all minds across time and space and hence helped to break the hold of individual subjectivity. Structural objectivity bypassed mechanical objectivity as it was reckoned that even the most carefully taken photographs could not yield results that were invariant from one observer to another. Just like structural objectivity, trained judgment was another response to the limitations of the empirical images and photographs used by mechanical objectivity. The new epistemic footprint was heralded by the transition from the understanding that, "objectivity should not be sacrificed to accuracy" (mechanical objectivity) to "accuracy should not sacrificed to objectivity" (trained judgment). The new epistemic virtue explicitly stated that: Automaticity of machines however sophisticated could not replace the professional practiced eye, namely trained judgment. Daston and Galison (2007) provide various examples of this change in the history of science such as Particle physics led by Luis Alvarez, Recognition by radiologists of errors in the naïve use of x-rays, Trained judgment was crucial in the Millikan-Ehrenhaft controversy with respect to the oil drop experiment and Martin Perl's discovery of the Tau Lepton (P. Galison, Email to author, November 17, 2015b). It is plausible to suggest that accumulation of experimental data in itself is not sufficient, and that the historical perspective shows that mechanical objectivity would approximate to the ideals of logical positivism and trained judgment to how science is actually done, namely "science in the making."

Understanding Objectivity in Research Reported in the Journal Science and Education (Springer) (Chap. 3). Based on a website search with the key word "objectivity," 131 articles in the 23-year period (1992–2014) referred to some form of objectivity and were classified according to the following criteria: Level I, traditional understanding of objectivity as found in science textbooks and positivist philosophers of science; Level II, a simple mention of objectivity as an academic/literary objective; Level III, problematic nature of objectivity is recognized, however, no mention is made of its changing/evolving nature; Level IV, an approximation to the evolving/changing nature of objectivity based on social and cultural aspects; Level V, a detailed historical reconstruction of the evolving nature of objectivity that recognized the role of the scientific community and its implications for science education. Results obtained showed the following distribution of the 131 articles evaluated: Level I = 5, Level II = 56, Level III = 58, Level IV = 10, and Level V = 2. Depending on the treatment of the subject 71 examples were selected to illustrate how the authors conceptualized objectivity and its evolution. Only 9% (12 out of 131) of the articles were considered to have an understanding of objectivity that approximated to its historical evolution. Four articles referred to the work of Daston and Galison on objectivity and only one mentioned "trained judgment." One article based on the work of Longino (explanatory plurality) reconciled the objectivity of science with its social and cultural construction (Level IV). Baconian notion of objectivity required the scientist to be neutral and detached from the research project. However, history of science shows that values play an important role in the development of science as data in and themselves do not determine how they are to be understood. Based on this background, one article endorsed Longino's pluralist approach to objectivity as it facilitates consensus formation through intersubjective assent (Level V). Gergen considers objectivity not to be static but rather differentiates it through two general categories of process and product. On the other hand, Daston and Galison refer to one stage in the history of objectivity as truth-to-nature. One article suggested a resemblance between the two approaches, as both recognize a stage in the history of objectivity that can be considered as an artifact of nature (Level V). The role played by observations is controversial in both science and science education. For example, it can be claimed that when two similar cameras take a picture of the same object they produce two identical images. In contrast, when two persons see the same experimental observations there are two different experiences. This suggests that pictures of the cameras are objective whereas the experiences of human beings are subjective. One article countered the argument by suggesting that it is precisely the role of science education to train people to be reliable observers (Level IV). Actually, this is explained by Daston and Galison as the reason why scientists started to question mechanical objectivity in the late nineteenth and early twentieth century and the underlying argument instead was precisely that of facilitating trained judgment in order to achieve consensus. Actual scientific practice is complex in which controversies based on the presuppositions of the protagonists play a crucial role. Although objectivity and open-mindedness are important attributes of science, these cannot be understood in the usual and naïve sense. One article suggested that rarely does a scientist commence research in the absence of presuppositions and thus objectivity consists not in denying preconceptions but rather in the ability to modify beliefs in the light of emerging evidence (Level IV).

School science generally presents the traditional view of science as objective and value free. One article emphasized that it is misleading to present a vision of science in which objectivity is considered to be an all or nothing thing. On the contrary, it is more realistic to suggest that objectivity is achieved in degrees (Level III). Communication and criticism are an important part of the scientific enterprise. Based on Longino, one article suggested that the peer-review process serves to enhance objectivity by decreasing the impact of individual scientists' subjectivity and thus facilitates scrutinized scientific knowledge (Level IV).

Understanding Objectivity in Research Reported in the Journal of Research in Science Teaching (Wiley-Blackwell) (Chap. 4). Based on a website search with the key word "objectivity," 110 articles in the 24-year period (1992–2015) referred to some form of objectivity and were classified according to the following criteria: Levels I–V (same as presented in Chap. 3). Results obtained showed the following distribution of the 110 articles evaluated: Level I = 4, Level II = 33, Level III = 68, Level IV = 5, and Level V = none. Depending on the treatment of the subject 49 examples were selected to illustrate how the authors conceptualized objectivity and its evolution. Only 5% (5 out of 110) of the articles were considered to have an understanding of objectivity that approximated to its historical evolution. None of the articles referred to the work of Daston and Galison on objectivity or mentioned "trained judgment." Traditional standards of educational research are based on positivist philosophy. One article reported that based on Guba and Lincoln's notion of trustworthiness traditional standards of internal and external validity, reliability and objectivity can be replaced by notions of credibility, transferability, dependability and triangulation of data sources (Level III). Based on the ideas of McLaren (a Marxist) and Harding (a feminist), one article explored the role of the unobtainable ideals of truth and objectivity and its consequences for school science. This may deny the students the opportunity to "learn the canon" or to "have access to the culture of power" and thus further oppress the marginalized community. The author later clarified that Harding does not assume that because a standpoint is articulated from the position of the oppressed that is necessarily the best position (Level III). The relationship between the production of knowledge and world views was explored by one of the articles. It was suggested that the dualism between objectivity and subjectivity leads to a conflict in the evolving nature of progress in science and that ignoring this duality may lead to the hegemony of objective knowledge in school science and consequent emphasis on rote learning (Level IV). In order to facilitate objectivity and researcher independence it is generally recommended in educational research that the researchers must maintain a distance between themselves and the subjects of their investigation. This prescription is, however, problematic as one article reported that in order to establish a mutually acceptable dialogue with the teacher in the classroom it is important to audit the process rather than the product (Level III). Teaching nature of science can follow two strategies, namely subjective factors such as theory ladenness, creativity and imagination and on the other hand objectivity. One of the articles, however, reported that extremes of subjectivity or objectivity are not desirable. During progress in science itself, the subjective and objective categories interact by means of communications (peer review) in the scientific community and the same can occur in the classroom (Level IV). Constructivism and relativism are controversial topics in science education. One of the articles has suggested that the role played by the scientific community in correcting knowledge claims is continuous and this represents open-mindedness and not relativism (Level IV). Historical and philosophical arguments have shown that both the epistemology of science and development of scientific theories are strongly dependent on social and cultural influences. With this perspective and based on the work of Collins, Fuller, and Holton, one article suggested that objectivity in its purest sense is never an option (Level III). One article referred to the support provided by science in the service of Enlightenment, when it introduced democratic and egalitarian notions that were resisted on political and religious considerations. Later, this fruitful relationship terminated as the objectivity of science provided the necessary evidence for the inferiority of women, homosexuals, the lower classes, the colonized and the enslaved (Level III). Teaching controversial topics such as evolutionary theory, in which both the participants and the researchers have prior epistemological views, can produce conflicting situations in the classroom. One article suggested that findings of such educational research can be seen as the two poles of the subjectivity–objectivity interface. In other words, the researcher based on his professional training in evolutionary biology thinks that he is being objective and at the same time in his interactions with the participants he is forced to understand their views and hence the need for subjective understanding (Level III).

Understanding Objectivity in Research Reported in Reference Works (Chap. 5). This chapter is based on the evaluation of research in two reference works: (a) International Handbook of Research in History, Philosophy, and Science Teaching (HPST); and (b) Encyclopedia of Science Education (ESE). Based on a website search with the key word "objectivity," 8 articles in the HPST and 12 articles in ESE referred to some form of objectivity and were classified according to the following criteria: Levels I–V (same as presented in Chap. 3). Results obtained showed the following distribution of the 20 articles evaluated in the two reference works: Level I = none, Level II = 10, Level III = 7, Level IV = 3, and Level V = none. Depending on the treatment of the subject, 20 examples were selected to illustrate how the authors conceptualized objectivity and its evolution. Only 15% (3 out of 20) of the articles were considered to have an understanding of objectivity that approximated to its historical evolution. One of the articles referred to the work of Daston and Galison on objectivity and none mentioned "trained judgment." One article referred to the difficulties involved in Harding's interpretation based on social and cultural factors that led her to conclude that claims of Western science to universality and objectivity should be rejected as illusions. This interpretation, however, is problematic as given the evolving nature of objectivity; it is absolute objectivity that remains as an illusion. In the context of feminist critiques of science, one article argued that objectivity is undermined if the correctness of a claim is taken to be what is endorsed by a privileged point of view. Consequently, for objectivity to be possible, no point of view can be globally privileged. There is some consensus that mathematical propositions are not empirically falsifiable and thus possess the absolute certainty of analytical statements or logical truths. One article has questioned this role of mathematical propositions as many advanced sciences are very much like mathematics in their conceptual apparatus, as can be illustrated with relativity and string theory. Consequently, as suggested by Popper the objectivity of mathematics is inseparably linked with its "criticizability." A major concern of science education is cognition and conceptual performance is highly rewarded. However, there is evidence that affect and cognition are inseparable and mutually constitutive. One article argued that inclusion of affect can open profound questions of objectivity and subjectivity and thus facilitate a history of science that is more in consonance with the practice of science. Constructivism in science education is a controversial topic. For example, radical constructivism was promoted by science educators who were dissatisfied with objectivism, namely scientific knowledge as an accurate depiction of physical reality. Similarly, the cornerstone concept of objectivity is reconceptualized as consensual agreement by scientific communities of practice, which comes quite close to what Daston and Galison (2007) have referred to as "trained judgment." The role of non-epistemic values such as ethical, social, and economic are being increasingly recognized as important for science education as these do not necessarily damage the progress, reliability and objectivity of science. In this context, one article posed the following question: if the ideal of value-free inquiry is flawed, what is to replace it? Based on Longino, a possible candidate would be "social value management," which incorporates non-epistemic values into science, subject to rigorous scrutiny of all possible perspectives.

Science at a Crossroads: Transgression versus Objectivity (Chap. 6). In this chapter I first explore the relationship between transgression and objectivity and then study the importance of Scanning tunneling microscope (STM) and Atomic force microscope (AFM) for chemical research (nanotechnology) and how these are presented in general chemistry textbooks. In order to understand scientific progress, Roald Hoffmann (2012), Nobel Laureate in chemistry, invokes the idea of "transgression of categorization" and Daston and Galison (2007) refer to violating the rules dictated by objectivity. When consulted, Hoffmann confirmed that the two concepts approximate to each other. Furthermore, both understand the transgression of objectivity in the context of Hacking's (1983) differentiation between "representation" and "intervention." Representation (fidelity to nature) has a long history that was variously understood as truth-to-nature, mechanical objectivity and trained judgment (for details see Chap. 2). On the other hand, presentation (intervention for Hacking) grew with nanotechnology in the late twentieth century (STM, AFM) and espouses object manipulation. Nanotechnology is not concerned about errors in our knowledge, nor if are dealing with real objects but rather with creating and manipulating to construct a new world of atom-sized objects. In this context it is plausible to suggest that at present progress in science is at a crossroads. This is particularly important for science educators as on the one hand they have to study, depict, and explain what actually exists (representation) and at the same time explore possibilities of what can be manipulated (presentation) to produce new products. In this quest, Hoffmann (2012) is emphatic that scientists have to go way beyond a prescribed procedure (scientific method). Based on this perspective, 60 general chemistry textbooks (published in USA) were evaluated on the following criteria: (1) Objectivity; (2) Scientific method; (3) STM; (4) AFM; and (5) From representation to presentation: Scientific progress at a crossroads. Textbooks were classified as satisfactory, mention or no mention. Percentages of textbooks that were considered to have a satisfactory presentation on the five criteria respectively were the following: 8, 18, 27, 12, and 25. This shows that understanding objectivity (Criterion 1, 8% satisfactory) was the most difficult for textbooks. In contrast, textbooks had a better understanding of STM (Criterion 3, 27% satisfactory) and scientific progress at a crossroads (Criterion 5, 25% satisfactory). It was found that understanding objectivity also leads to a better understanding of scientific method. One textbook referred to the problematic nature of objectivity as experiments often have some level of uncertainty, spurious and contradictory data can be collected leading to the conclusion that the original hypothesis itself needs changes. Under such circumstances it is difficult for the scientists to remain objective. Some textbooks present the traditional scientific method and at the same time recognize the importance of doubts, conflicts, skepticism, clashes of personalities, and even revolutions of perception in actual historical episodes. This is an innovative step and helps in understanding "science in the making." Some textbooks explicitly referred to the difference between the images of an optical microscope and the computer-generated images produced by STM, based on wave-mechanical properties of surface electrons and do not provide information about the internal structure of atoms. Elaboration of nanoscale electronic components was referred to by some textbooks as a dream come true, and that we now have a third form of carbon (buckminsterfullerene) besides those mentioned in the periodic table, namely graphite and diamond. The new form of carbon is initially as soft as graphite, but when compressed by 30% it becomes harder than diamond. Furthermore, when this pressure is removed the solid springs back to its original volume. Such discoveries can help students to understand that scientific progress is at a crossroads.

Conclusion: Understanding the Elusive Nature of Objectivity (Chap. 7). An evaluation of research reported in this book shows the problematic nature of understanding some of the universal values associated with objectivity, such as certainty, value neutral observations, facts, infallibility, and truth of scientific theories and laws. These results provide a detailed account (over a period of almost 25 years) of how the science education research community conceptualizes the difficulties involved in accepting objectivity as an unquestioned epistemic virtue of the scientific enterprise. Analyses of general chemistry textbooks are used to introduce the idea of "transgression of objectivity" and that scientific progress (nanotechnology) is at a crossroads. Given the importance of objectivity/subjectivity dichotomy in science education, it is plausible to suggest that objectivity has become an opiate of the academic. Although, achievement of objectivity in actual scientific practice is a myth, it still remains a powerful and useful idea. It seems that more work needs to be done in order to facilitate a transition toward a more nuanced understanding of objectivity and eventually the dynamics of scientific progress.

References

Abd-El-Khalick, F. (2012). Examining the sources for our understandings about science: enduring conflations and critical issues in research on nature of science in science education. *International Journal of Science Education, 34*(3), 353–374.

Agazzi, E. (2014). *Scientific objectivity and its contexts*. Heidelberg: Springer.

American Association for the Advancement of Science, AAAS. (1993). *Benchmarks for science literacy: project 2061*. Washington: Oxford University Press.

Australian Curriculum and Reporting Authority, ACARA. (2015). *Australian curriculum: science F-10*. Sydney: Commonwealth of Australia.

Berger, J. O., & Berry, D. A. (1988). Statistical analysis and the illusion of objectivity. *American Scientist, 76*(2), 159–165.

Campbell, D. T. (1988a). Can we be scientific in applied social science? In E. S. Overman (Ed.), *Methodology and epistemology for social science* (pp. 315–333). Chicago: University of Chicago Press. (first published in 1984).

Campbell, D. T. (1988b). The experimenting society. In E. S. Overman (Ed.), *Methodology and epistemology for social science* (pp. 290–314). Chicago: University of Chicago Press.

Cawthron, E. R., & Rowell, J. A. (1978). Epistemology and science education. *Studies in Science Education, 5*, 51–59.

Chang, Y.-H., Chang, C.-Y., & Tseng, Y.-H. (2010). Trends of science education research: an automatic content analysis. *Journal of Science Education and Technology, 19*, 315–331.

Council of Ministers of Education, CMEC. (1997). *Common framework of science learning outcomes K to 12: Pan-Canadian protocol for collaboration on school curriculum*. Toronto: Council of Ministers of Education.

Daston, L., & Galison, P. L. (1992). The image of objectivity. *Representations, 40*, 81–128. (special issue: Seeing Science).

Daston, L., & Galison, P. (2007). *Objectivity*. New York: Zone Books.

Deng, F., Chai, C. S., Tsai, C.-C., & Lin, T.-J. (2014). Assessing South China (Guangzhou) high school students' views on nature of science: a validation study. *Science & Education, 23*, 843–863.

Galison, P. (2015a). The journalist the scientist and objectivity. In F. Padovani, A. Richardson & J. Y. Tsou (Eds.), *Objectivity in science. Boston Studies in the Philosophy and History of Science*. Dordrecht: Springer.

Giere, R. N. (1999). *Science without laws*. Chicago: University of Chicago Press.

Giere, R. N. (2006a). Perspectival pluralism. In S. H. Kellert, H. E. Longino & C. K. Waters (Eds.), *Scientific pluralism* (pp. 26–41). Minneapolis: University of Minnesota Press.

Giere, R. N. (2006b). *Scientific perspectivism*. Chicago: University of Chicago Press.

Giere, R. N. (2010). Naturalism. In S. Psillos & M. Curd (Eds.), *The Routledge companion to philosophy of science* (pp. 213–223). London: Routledge.

Gould, S. J. (1995). *Dinosaur in a haystack: reflections in natural history*. New York: Crown Trade Paperbacks.

Hacking, I. (1983). *Representing and intervening*. Cambridge: Cambridge University Press.

Harding, S. (2015). *Objectivity and diversity: another logic of scientific research*. Chicago: University of Chicago Press.

Hodson, D., & Wong, S. L. (2014). From the horse's mouth: why scientists' views are crucial to nature of science understanding. *International Journal of Science Education, 36*(16), 2639–2665.

Hoffmann, R. (2012). In J. Kovac & M. Weisberg (Eds.), *Roald Hoffmann on the philosophy, art, and science of chemistry*. New York: Oxford University Press.

Holton, G. (1969). Einstein and the 'crucial' experiment. *American Journal of Physics, 37*, 968–982.

Holton, G. (1978a). Subelectrons, presuppositions, and the Millikan-Ehrenhaft dispute. *Historical Studies in the Physical Sciences, 9*, 161–224.

Holton, G. (1978b). *The scientific imagination: case studies*. Cambridge: Cambridge University Press.

Holton, G. (2014a). The neglected mandate: teaching science as part of our culture. *Science & Education, 23*, 1875–1877.

Johnson, R. B., & Onwuegbuzie, A. J. (2004). Mixed methods research: a research paradigm whose time has come. *Educational Researcher, 33*, 14–26.

Kuhn, T. (1970). *The structure of scientific revolutions*. Chicago: University of Chicago Press. (2nd ed.).

Kuhn, T. (1977). Objectivity, value judgment, and theory choice. In T. Kuhn (Ed.), *The essential tension* (pp. 320–339). Chicago: University of Chicago Press. (first presented as a Lecture at Furman University in 1973).

Lakatos, I. (1970). Falsification and the methodology of scientific research programs. In I. Lakatos & A. Musgrave (eds.), *Criticism and the growth of knowledge* (pp. 91–195). Cambridge: Cambridge University Press.

Lederman, N. G. (2007). Nature of science: past, present, and future. In S. K. Abell & N. G. Lederman (Eds.), *Handbook of research on science education* (pp. 831–879). Mahwah: Lawrence Erlbaum.

Lederman, N. G., Abd-El-Khalick, F., Bell, R. L., & Schwartz, R. (2002). Views of nature of science questionnaire: toward valid and meaningful assessment of learners' conceptions of nature of science. *Journal of Research in Science Teaching, 39*, 497–521.

Machamer, P., Pera, M., & Baltas, A. (2000). Scientific controversies: an introduction. In P. Machamer, M. Pera & A. Baltas (Eds.), *Scientific controversies: philosophical and historical perspectives* (pp. 3–17). New York: Oxford University Press.

Machamer, P., & Wolters, G. (2004). Introduction: science, values and objectivity. In P. Machamer & G. Wolters (Eds.), *Science, values and objectivity* (pp. 1–13). Pittsburgh: University of Pittsburgh Press.

McComas, W. F., Clough, M. P., & Almazroa, H. (1998). The role and character of the nature of science in science education. In W. F. McComas (Ed.), *The nature of science in science education: rationales and strategies* (pp. 3–40). Dordrecht: Kluwer.

Medawar, P. B. (1969). *Induction and intuition in scientific thought*. Philadelphia: American Philosophical Society.

Nagel, E. (1961). *The structure of science: problems in the logic of scientific explanation*. New York: Harcourt, Brace & World, Inc..

National Research Council, NRC. (2013). *Next Generation Science Standards (NGSS)*. Washington: National Academies Press. (http://www.nextgenscience.org).

Niaz, M. (2009). *Critical appraisal of physical science as a human enterprise: dynamics of scientific progress*. Dordrecht: Springer.

Niaz, M. (2012). *From 'Science in the Making' to understanding the nature of science: an overview for science educators*. New York: Routledge.

Niaz, M. (2016). *Chemistry education and contributions from history and philosophy of science*. Dordrecht: Springer.

Phillips, D. C., & Burbules, N. C. (2000). *Postpositivism and educational research*. New York: Rowman & Littlefield.

Resnik, D. B. (2010). Ethics of science. In S. Psillos & M. Curd (Eds.), *The Routledge companion to philosophy of science* (pp. 149–158). London: Routledge.

Schwab, J. J. (1974). The concept of the structure of a discipline. In E. W. Eisner & E. Vallance (Eds.), *Conflicting conceptions of curriculum* (pp. 162–175). Berkeley: McCutchan Publishing Corp..

Shapin, S. (1996). *The scientific revolution*. Chicago: University of Chicago Press.

Smith, M. U., & Scharmann, L. C. (2008). A multi-year program developing an explicit reflective pedagogy for teaching pre-service teachers the nature of science by ostention. *Science & Education, 17*, 219–248.

Smith, M. U., Siegel, H., & McInerney, J. D. (1995). Foundational issues in evolution education. *Science & Education, 4*(1), 23–46.

Smolicz, J. J., & Nunan, E. E. (1975). The philosophical and sociological foundations of science education: the demythologizing of school science. *Studies in Science Education, 2,* 101–143.

Tsou, J. Y., Richardson, A., & Padovani, F. (2015). Introduction. In F. Padovani, A. Richardson & J. Y. Tsou (Eds.), *Objectivity in science. Boston Studies in the Philosophy and History of Science.* Dordrecht: Springer.

Vesterinen, V.-M., & Aksela, M. (2013). Design of chemistry teacher education course on nature of science. *Science & Education, 22*(9), 2193–2225.

Wong, S. L., & Hodson, D. (2009). From the horse's mouth: what scientists say about scientific investigation and scientific knowledge. *Science Education, 93,* 109–130.

Ziman, J. (2000). *Real science: what it is, and what it means.* New York: Cambridge University Press.

Chapter 2
Objectivity in the Making

2.1 Theoretical Framework

The theoretical framework of the studies in this book is primarily based on an examination of the evolving forms of scientific judgment (including objectivity) in the history of science as suggested by Daston and Galison (1992, 2007). The subject of objectivity, its precursors and followers, is important not only for science but also for science education. In order to facilitate understanding it would be of interest to consider the following markers (even perhaps brain-storming ideas) in the work of Daston and Galison (2007):

- The history of objectivity is nothing less than the history of science itself (p. 34).
- To study objectivity in shirt-sleeves is to watch objectivity in the making (p. 52).
- There is no objectivity without subjectivity to suppress, and vice versa (p. 33).
- Objectivity and subjectivity define each other, like left and right (p. 36).
- What are the epistemological pretensions of objectivity? (p. 51).
- Objectivity is assumed to be abstract, timeless, and monolithic (p. 51).
- Do "objective methods" guarantee the truth of a finding? (p. 51).
- Truth did not lie on the visible surface of the world (p. 185).
- Objectivity was a desire, a passionate commitment to suppress the will, a drive to let the visible world emerge on the plate without intervention (p. 143).

In order to facilitate an understanding of the principal theses of their book, Daston and Galison (2007) present three images as a historical series:

1. *Truth-to-nature. Campanula foliis hastatis dentatis* by Carolus Linnaeus, *Hortus Cliffortianus*, published in Amsterdam in 1737. An illustration of a landmark botanical work (flower) aimed to portray the characteristic, the essential, the universal, and the typical: truth-to-nature (p. 20).
2. *Mechanical objectivity.* Snowflake by Gustav Hellmann, *Schneekrystalle: Beobachtungen und Studien*, published in Berlin in 1893. An individual

© Springer International Publishing AG 2018

M. Niaz, *Evolving Nature of Objectivity in the History of Science and its Implications for Science Education*, Contemporary Trends and Issues in Science Education,
DOI 10.1007/978-3-319-67726-2_2

snowflake with all its peculiarities and asymmetries in an attempt to capture nature with as little human intervention as possible: mechanical objectivity (p. 20).

3. *Trained judgment.* Solar magnetogram in the *Atlas of Solar Magnetic Fields*, published in Washington, DC, in 1967. This image of the magnetic field of the sun mixed the output of sophisticated equipment with a "subjective" smoothing of the data—the authors deemed this intervention necessary to remove instrumental artifacts: trained judgment (p. 21).

In this historical sequence each successive stage presupposes and builds upon, as well as reacts to, the earlier ones. Truth-to-nature (universal and the typical) was a precondition for mechanical objectivity (presenting nature without intervention), and mechanical objectivity was a precondition for trained judgment (subjective intervention of the data). According to Daston and Galison (2007), behind the flower, the snowflake, the solar magnetogram stand not only the scientist who sees and the artist who depicts, but also a certain collective way of knowing (p. 53). Nature, knowledge, and knower intersect in these images, and thus the world becomes intelligible. It is precisely this intersection in the history of science that has led to different forms of understanding scientific judgment. Truth-to-nature, mechanical objectivity, and trained judgment are presented in detail in the next sections.

According to Daston and Galison (2007), what is knowledge and how it is attained can be understood by the following sequence of historical events and practices, which helped to understand objectivity in the making: truth-to-nature (eighteenth century), mechanical objectivity (nineteenth century), structural objectivity (late nineteenth century), and trained judgment (twentieth century). These authors have based their work on scientific atlas images as these have been central to scientific practice in different periods and across disciplines such as anatomy, physics, meteorology, and embryology, among others.

2.1.1 Truth-to-Nature

Truth-to-nature refers to science before objectivity, as practiced by Enlightenment naturalists in the eighteenth century, based on: selecting, comparing, judging, and generalizing. According to Daston and Galison (2007), in truth-to-nature it was important that the naturalist be steeped in but not enslaved to nature as it appeared (p. 59). To illustrate, Daston and Galison (2007) provide the following example from what Johann Wolfgang von Goethe wrote in 1798 with respect to his research in morphology and optics by emphasizing that the human mind must fix the empirically variable, exclude the accidental, eliminate the impure, unravel the tangled, and eventually discover the unknown. For Goethe's contribution also see Fara (2009, p. 257). Scientists who followed truth-to-nature were *looking for the idea in the observation and not the raw observation itself.* Objects that were depicted in the atlases did not represent particulars but universals, that is idealization. The work of the Swedish naturalist Carolus Linnaeus and Goethe are good examples of such endeavors. In 1737, Linnaeus published a flora of the plants cultivated in the well-stocked garden of George Clifford, an Amsterdam banker and director of the Dutch East India Company. Linnaeus advised Botanists to concentrate on characters that are constant, certain and organic and not be distracted by irrelevant details of a plant's appearance. In short his ways of describing, depicting, and classifying plants were openly and even aggressively selective, and this precisely constituted truth-to-nature. Linnaeus rejected that the scientific knowledge most worth seeking was that which depended least on the personal traits of the researcher. The tenets of objectivity, as they were formulated in the mid-nineteenth century, would have contradicted Linnaeus's sense of the scientific endeavor and he would have dismissed as irresponsible the suggestion that scientific facts should be conveyed without the mediation of the scientist. Precisely, for this reason the followers of mechanical objectivity in the nineteenth century considered drawings of Linnaeus as subjective distortions. Mechanical objectivity, however, did not extinguish truth-to-nature, but rather collided and coexisted.

2.1.2 Mechanical Objectivity

Scientists and atlas makers committed to mechanical objectivity were particularly critical of those who followed truth-to-nature and considered it as subjective distortion based on selection, synthesis, and idealization. Those following mechanical objectivity called for objective photographs to supplement, correct, or even replace the subjective drawings produced by those who followed truth-to-nature. By the late nineteenth century, although mechanical objectivity did not drive out truth-to-nature, it became firmly established as a guide for scientific representation across a wide range of disciplines (Daston & Galison, 2007, p. 111).

The controversy between two histologists in the late nineteenth and early twentieth century, Santiago Ramón y Cajal from Spain and Camillo Golgi from Italy, is quite representative of the issues involved in mechanical objectivity and truth-to-nature, respectively (Daston & Galison, 2007, pp. 115–120). Golgi claimed that his drawings and descriptions of the cerebrum, cerebellum, spinal cord, and hippocampus were "exactly prepared according to nature," namely examining the microscopic specimen and then modifying the figures to make them look less complicated than in nature. This precisely represented truth-to-nature, based on Golgi's theory of interstitial nerve nets. On the contrary, Ramón y Cajal based on his neuron doctrine considered that Golgi by simplifying nature was not being objective. Objectivity was the central issue in the debate: Cajal defended his undistorted sight and charged Golgi of having intervened deliberately in accordance with his theoretical predilections. Ramón y Cajal (1989) considered the joint award along with Golgi of the Nobel Prize as an injustice: "What a cruel irony of fate to pair, like Siamese twins united by the shoulders, scientific adversaries of such contrasting character!" (p. 553). Although the alteration of the image easily led to the dreaded subjectivity of interpretation, Daston and Galison (2007) have raised a very pertinent question: Could Golgi, Cajal, or, for that matter, anyone else dispense fully with all intervention, and responded in the negative. Everyone recognized that and thus mechanical objectivity remained an always-receding ideal, and thus never fully obtainable. The Cajal-Golgi battle may remind readers of the "battle over the electron" (early twentieth century), with the difference that both got the Nobel Prize for Physiology or Medicine in 1906, whereas in the latter case Robert Millikan got the Physics Nobel Prize in 1923 and Felix Ehrenhaft was ignored (cf. Holton, 1978a, b; Niaz, 2005). The Millikan-Ehrenhaft controversy will be discussed later in this chapter.

According to Daston and Galison (2007, p.187), the photograph became the emblem for all aspects of noninterventionist objectivity, and this was primarily due to the fact that the camera apparently eliminated human agency. For Cajal and others with similar thinking nonintervention lay at the heart of mechanical objectivity.

Photography had its own problems with respect to reflecting the object objectively, and this was recognized early by Richard Neuhauss (1898), an expert on photomicrography, as too much light or too little light changed the details in a photograph. The light-sensitive photographic plate copies everything even if something does not belong to the object, such as impurities, diffraction edges, dust particles, plate defects, and many other artifacts. After working for 40 years in the service of scientific photography Neuhauss became convinced that mechanical objectivity, based on automaticity and noninterference by the scientist, was difficult to achieve (cf. Daston & Galison, 2007, pp. 187–189).

According to Daston and Galison (2007, pp. 197–198), objectivity and subjectivity are as inseparable as concave and convex, and one defines the other. The emergence of scientific objectivity in the mid-nineteenth century necessarily goes hand in glove with the emergence of scientific subjectivity. The extraordinary measures of mechanical objectivity were invented and mobilized to combat the enemy, namely subjectivity. In the early twentieth century, many scientists became convinced that subjectivity was difficult to separate from objectivity, and some became skeptical of engravings, drawings, and photographs and instead started to look in the domain of mathematics and logic.

2.1.3 Structural Objectivity

Just as scientists and atlas makers were busy in the mid-nineteenth century in adopting mechanical objectivity, voices of dissent also started to appear especially with respect to the distinction between observation (astronomer in the observatory) and experiment (chemist in the lab). One such voice was that of Claude Bernard (1865) working in experimental medicine, who cautioned with respect to the distinction between passive observation and active experimentation:

> Yes, no doubt, the experimenter forces nature to unveil herself, attacking her and posing questions in all directions; but he must never answer for her nor listen incompletely to her answers by taking from the experiment only the part that favors or confirms the

hypothesis …. One could distinguish and separate the experimenter into he who plans and institutes the experiment from he who executes it and registers the results (p. 53).

This was written in the middle of the nineteenth century, and its meaning may be elusive for present-day students of history of science. The following commentary by Daston and Galison (2007) helps to understand the real import: "One and the same scientist had somehow both to be *speculative and bold in designing an experiment* to pry answers out of nature *and* to obtain the results passively, as if in ignorance of the hypothesis the experiment aimed to test. The scientist was both inquisitor and confessor to nature" (p. 243, italics added). Interestingly, according to Holton (1978a, b, pp. 184–185) Millikan did not design the oil drop experiment but rather discovered it. In other words, it was the electron theory which suggested the existence of the elementary electrical charge and hence the need for its experimental determination. Similarly, Martin Perl in his search for fractional charges (quarks) in the late twentieth century has provided the following advice: "Choices in the *design of speculative experiments* usually cannot be made simply on the basis of pure reason. The experimenter usually has to base his/her decision partly on what feels right, partly on what technology they like, and partly on what aspects of the speculations they like" (Perl & Lee, 1997, p. 699, italics added). Indeed, the dilemma faced by scientists in the mid-nineteenth and late twentieth century was quite similar and required the design of speculative experiments. There is, however, one important difference as most scientists are aware that the path from experiments to results and their interpretation is laden with controversies. Interestingly, both Millikan and Perl conflated the roles of "inquisitor" and "confessor" to nature.

Scientists and atlas makers following mechanical objectivity were facing two difficulties: (a) It demanded that the scientific self, split into active experimenter and passive observer; and (b) It required a universal working object to be extracted from a particular specimen. These contradictions led some to adopt structural objectivity and others trained judgment as alternatives.

In the late nineteenth- and early twentieth-century many scientists adopted a version of objectivity grounded in structures rather than images. Structures could be communicated to all minds across time and space and hence helped to break the hold of individual subjectivity. Structural objectivity lay not in the observable facts of mechanical objectivity but only in final invariants of experience, such as electrons, the ether. Frege, Carnap, Poincaré, Schlick, Russell, and Margenau, all followers of structural objectivity longed for a world that could be communicated and not just experiences. At this stage it would be interesting to consider if Einstein was a structural objectivist? Daston and Galison (2007) responded: yes and no (p. 305). Einstein considered the characterization of objectivity through invariant structures as far too narrow. Again, for Einstein to identify mathematical-physical structure with objectivity was far too broad. Furthermore, Einstein considered that time could be defined objectively only alongside space and thus took special relativity to shatter the objectivity that seemed to characterize time by itself (p. 302).

Structural objectivity emphasized structural relations rather than objects *per se* (Daston & Galison, 2007). Rejection of mechanical objectivity led to an intensification of objectivity on another scale and convinced structural objectivists that even the most carefully taken photograph would never yield results truly invariant from one observer to another (p. 317).

2.1.4   Trained Judgment

Early in the twentieth century, scientists and atlas makers came to see the limitations of mechanical objectivity and the need for going beyond by employing trained judgment based on an interpretative vision of the scientific enterprise. Just like structural objectivity, trained judgment was another response to the limitations of the empirical images and photographs used by mechanical objectivity. Within a historical perspective scientists following truth-to-nature (idealized objects) were led to mechanical objectivity (actual images and photographs), which in turn led to structural objectivity (relational invariants) and finally came trained judgment, through interpreted images (based on "trained" or "seeing" eye). Of course, this historical transition does not mean that each replaced the other, that is instead of supplanting, these different forms of understanding science supplemented each other. Within the historical perspective formulated by Daston and Galison (2007), the different forms of objectivity represent alternatives that can be supported by groups with different philosophical orientations.

Daston and Galison (2007) have provided a detailed overview of how trained judgment came to be an important part of scientific understanding and following are some of the examples provided by them:

(a) In the preface of their celebrated *Atlas of Electroencephalography*, Gibbs and Gibbs (1941) noted: "this book has been written in the hope that it will help the reader to see at a glance what it has taken others many hours to find, that it will help to train his eye so that he can arrive at diagnoses from *subjective* criteria" (Preface, n.p.). Interestingly, after citing this, Daston and Galison (2007) commented: "Could it be that Gibbs and Gibbs simply did not understand the way 'objective' and 'subjective' had been deployed by the mechanical objectivists of the previous hundred years? Could they be 'talking past' those who deplored the subjective? No, the Gibbs understood full well the pictorial practice of mechanical objectivity. And they emphatically rejected it ..." (p. 322). Furthermore, it is important to note that the Gibbs were the pioneers and considered as world's experts in the new and sophisticated electroencephalogram. Ten years later in the new edition of the *Atlas of Encephalography*, Gibbs and Gibbs (1951) went even beyond and stated: "Experimentation with wave counts ... and with frequency analysis of the electroencephalogram ... indicate[s] that no objective index can equal the accuracy of subjective evaluation ... Accuracy should not be sacrificed to objectivity ..." (p. 112). By any standard, the last statement in the above quote is thought provoking and this led Daston and Galison (2007) to comment: "This astonishing statement—astonishing from the perspective of mechanical objectivity—is the epistemic footprint of the new, mid-twentieth century regime of the interpreted image. How different this is from the reverse formulation of mechanical objectivity: that objectivity should not be sacrificed to accuracy" (p. 324). This transition from "objectivity should not be sacrificed to accuracy" (mechanical objectivity) to "accuracy should not be sacrificed to objectivity" (trained judgment), shows the stark difference between the dominance of mechanical objectivity for almost 100 years (1830s to 1930s) and the new "epistemic footprint" based on trained judgment. This led Daston and Galison (2007, p. 324) to understand the new epistemic footprint in the following terms: for advocates of rigorously trained judgment (e.g., Gibbs and Gibbs), the "autographic" automaticity of machines however sophisticated, could not replace the professional, practiced eye.

(b) In the 1960s, Luis Alvarez presided over a vast team of physicists, engineers, programmers, and scanners in what was the most highly instrumented particle physics laboratory in the world. In the training guide for all scanners it was pointed out that scanning techniques were approximate and that track density information was not foolproof. Actually, Alvarez was quite convinced that human beings have remarkable inherent scanning abilities (eyeballing) that are better than what can be built into a computer (Daston & Galison, 2007, p. 330).

(c) Radiologists have recognized the importance of errors in the naïve use of x-rays. In such images it is difficult to distinguish between variations within the bounds of the "normal" and variations that transgress normalcy and enter the pathological (Daston & Galison, 2007, p. 309). Keats (1973) in his *Atlas* of x-rays recognized that: "The proof of the validity of the material presented is largely subjective, based on personal experience and on the published work of others. It consists largely of having seen the entity many times and of being secure in the knowledge that time has proved the innocence of the lesions" (p. vii).

(d) Determination of the elementary electrical charge, the electron (e), has been the subject of considerable controversy between Robert Millikan (University of Chicago) and Felix Ehrenhaft (University of Vienna) that lasted for many years (around 1910–1923, when Millikan was awarded the Physics Nobel Prize). Both physicists had very similar experimental data and still Millikan postulated the existence of a universal charged particle (the electron) and Ehrenhaft postulated the existence of subelectrons based on fractional charges. The experiment and the empirical data became important in the light of a heuristic principle, namely the corroboration of the atomic nature of electricity based on a universal charged particle. Almost 55 years later, Holton (1978a, b) added a new dimension to the controversy with his discovery of Millikan's two laboratory notebooks at the California Institute of Technology, Pasadena. In these notebooks, Holton found data from 140 drops, but the published article (Millikan, 1913) reported results from only 58 drops. What happened to the other 82 drops? It seems that Millikan made a rough calculation for the value of e as soon as the data for the times of descent/ascent of the oil drops started coming in and ignored any drop that did not give the value of e that he expected according to his presuppositions (for details see Niaz, 2005). More recently, Holton (Email to author, August 3, 2014b) has clarified that, "So even if Millikan had included *all* drops and yet had come out with the same result, the error bar of Millikan's final result would not have been remarkably small, but large—the very thing Millikan did not like" (p. 1, italics in the original). Interestingly, according to Daston and Galison (2007, p. 478) the exercise of scientific judgment in the Millikan-Ehrenhaft controversy can be considered as an example of trained judgment. Galison (Email to author, November 17, 2015b) explicitly endorsed that trained judgment was fundamental for Millikan's work. Interestingly, the Millikan-Ehrenhaft controversy has been almost completely ignored in the following science textbooks: general chemistry textbooks published in USA (Niaz, 2000) and Turkey (Niaz & Coştu, 2013), general physics textbooks published in USA (Rodríguez & Niaz, 2004a).

Besides these examples, in my opinion, the following historical episodes can also be considered as examples of trained judgment:

1. In the early twentieth century, both J.J. Thomson and E. Rutherford obtained very similar experimental results based on the scattering of alpha particles. Based on these empirical findings, Thomson propounded the hypothesis of compound scattering (multitude of small scatterings) whereas Rutherford propounded the hypothesis of single scattering, in order to explain the large angle deflections of alpha particles (cf. Heilbron, 1981; Niaz, 2009; Wilson, 1983). At this stage a science student may ask: if experimental data alone leads to objectivity in science why did Thomson and Rutherford put forward two different atomic models? The scientific community considered the hypotheses put forward by both scientists and ultimately Rutherford's hypothesis triumphed (despite Thomson's opposition) not because of the experimental data but for the following reasons: (a) A total deflection greater than 90° in traversing the gold foil would have only one chance in 10^{3500} of occurring; and (b) Therefore, large angle deflections as a result of many single deflections in the same direction were very improbable. This clearly shows the need for interpretation of the experimental data and that requires trained judgment. This episode is all the more interesting as both Thomson and Rutherford were well known to each other (this contrasts with Millikan and Ehrenhaft who lived in different countries and even when Ehrenhaft immigrated to the USA, I am not sure if they ever met). However, by the time Ehrenhaft immigrated the controversy was almost over. The Thomson-Rutherford controversy is difficult for students to understand as they think that both could have met over dinner and resolved their differences. This experiment has been the subject of considerable research in the science education literature. For example, Niaz (1998) reported that none of the general chemistry textbooks published in USA, referred to the Thomson-Rutherford controversy. Similarly, the following textbooks almost completely ignored the controversy: general chemistry textbooks published in Turkey (Niaz & Coştu, 2009), general physics textbooks published in USA (Rodríguez & Niaz, 2004b) and South Korea (Niaz, Kwon, Kim, & Lee, 2013). It is plausible to suggest that inclusion of historical controversies in the science classroom can facilitate a better understanding of objectivity.

2. Observational data based on the bending of light in the 1919 eclipse experiments (Dyson, Eddington, & Davidson, 1920) showed that: of the three sources of

observational data the Principe (West coast of Africa) astrographic photographs were the worst, perhaps in part due to the cloudy weather. Data from the two Sobral (East coast of Brazil) telescopes were also affected by clouds and provided two different sets of measurements. The mean of the deflection from the Sobral 4-inch telescope was significantly higher than Einstein's prediction (1.98″ vs 1.75″), whereas the mean of the deflection from the Sobral astrographic telescope came quite close to the Newtonian prediction. Under these circumstances Dyson and Eddington adopted the following strategy: Deflection from Sobral 4-inch and Principe photographs was considered acceptable (both being close to the value predicted by Einstein), whereas the deflection from the Sobral astrographic telescope was rejected on the grounds of systematic errors. According to Earman and Glymour (1980) given the importance of the issues at stake, the results should have been unequivocal, which they were not and could be held to confirm Einstein's theory, only if many of the measurements were ignored. To understand the issues better let us consider the following scenario: Eddington and Dyson are not aware of Einstein's General Theory of Relativity and particularly of the prediction that starlight near the sun would bend. Under these circumstances experimental evidence from all three sources (Sobral and Principe) would have been extremely uncertain, equivocal and difficult to interpret (for details see Niaz, 2009, Chap. 9, pp. 127–137). At this stage, it is important to note that on reading the above account, Galison (Email to author, November 17, 2015b) pointed out that the equivocal nature of the Edington data is often exaggerated, and instead he follows the views of Stanley (2007).

3. Martin Perl and colleagues working at the Stanford Linear Accelerator Center (SLAC), presented experimental evidence for having discovered the Tau Lepton in 1975. Of the 126 particle-pair events reported, at least 24 could be attributed to electron-muon events, which was the strongest evidence at that time for the Tau Lepton. Perl and colleagues, however, were not yet prepared to claim that they had found a new charged lepton, and in order to accentuate their uncertainty they denoted the new particle with U for "unknown" in some of their 1975–1977 papers (for details see Niaz, 2012, Chap. 7, pp. 196–204). Perl was awarded the Physics Nobel Prize in 1995 for his discovery of the Tau Lepton. Later recalling his experience Perl (2004) attributed his success (besides a strong belief in his presuppositions) to: "*I had smart, resourceful, and patient research companions. I think these are the elements that should be present in speculative experimental work:a broad general plan, specific research methods, new technology, and first-class research companions. Of course, the element of luck will in the end be dominant*" (pp. 418–419, original italics, underline added). The reference to "speculative experimental work" can indeed be a source of curiosity for some science students. Once again it can be seen that besides the experimental data, scientists need something else, namely trained judgment. Galison (Email to author, November 17, 2015b) explicitly endorsed that trained judgment was fundamental for Perl's discovery of the Tau Lepton.

4. History of quantum mechanics helps to understand the relationship between historical contingency and the evolving nature of objectivity. According to

James Cushing (1989): "Science is an historical entity whose practice, methods and goals are *contingent*. There may not be *a* rationality which is the hallmark or the essence of science" (p. 2, original italics. In an endnote Cushing explains what he means by contingent, "I simply mean not fixed by logic or necessity," p. 20). Around 1927, besides the Copenhagen interpretation of quantum mechanics, there were two rival interpretations, namely Schrödinger's wave picture and de Broglie's pilot-wave model—a precursor to Bohm's theory of hidden variables. According to Cushing (1996): "Given the *presumed objectivity and impartiality* of the scientific enterprise, one might expect that such an interpretation (Bohm's) would be given serious consideration by the community of theoretical physicists. However, it was basically ignored, rather than either studied or rebutted. Just as what are often termed 'external' factors had played a key role in establishing the Copenhagen hegemony, so they once again contributed to keeping this competitor from the field. That a generation of physicists had been educated in the Copenhagen dogma made it all the more difficult for Bohm's theory" (p. 13, italics added).

5. According to Gavroglu and Simões (2012), the rivalry between the valence bond and molecular orbital theories in chemistry can also be considered as an example of historical contingency. Even today, after almost 70 years, the two theories continue to be rivals and according to Hoffmann et al. (2003): "Discarding any one of the two theories undermines the intellectual heritage of chemistry" (p. 755). For details with respect to how the contingency thesis helps to understand the development of the two theories (valence bond and molecular orbital) and its presentation in general chemistry textbooks, see Niaz (2016, Chap. 6, pp. 143–158). In other words, as both theories were supported by experimental evidence, it was the intervention of the scientific community (trained judgment) that helped to resolve the controversy.

6. Holton (1969) has referred to the "experimenticist fallacy," frequently found in the scientific endeavor and in science curricula and textbooks. Campbell (1988a), a methodologist, has called attention to the pitfalls involved in experimenticism: "The objectivity of physical science does not come from turning over the running of experiments to people who could not care less about the outcome, nor from having a separate staff to read the meters. It comes from a social process that can be called competitive cross-validation … and from the fact that there are many independent decision makers capable of rerunning an experiment, at least in a theoretically essential form" (p. 324). Interestingly, Daston and Galison (2007) have recognized that the combining of automatic procedures with trained judgment, and the increasing reliance on pattern-recognition capabilities of a trained, educated audience is widespread in domains as diverse as geology, particle physics, and astronomy (p. 329). Similarities in the views of Holton and Campbell on the one hand and those of Daston and Galison are striking.

In all the examples discussed above the accumulation of experimental data in itself was not sufficient, but instead the role of the scientific community was crucial and that illustrates various aspects of "trained judgment." This clearly shows the elusive nature of objectivity in scientific progress. At this stage it is important to note (as suggested by one of the reviewers of this book) that most of the examples I have dealt with are from the physical sciences. However, given the importance of understanding objectivity, I am sure researchers in science education would explore other areas of expertise (e.g., biology, medicine, earth science).

Understanding and teaching about objectivity is perhaps one of the most difficult and controversial topic in the science education curriculum. To grant objectivity a history, Daston and Galison (2007) have explored the complexities of the issues involved in the following terms:

> The opposition between science as a set of rules and algorithms rigidly followed versus science as tacit knowledge (Michael Polanyi with a heavy dose of the later Ludwig Wittgenstein) no longer looks like the confrontation between an official ideology of scientists as supported by logical positivist philosophers versus the facts about how science is actually done as discovered by sociologists and historians. Instead, both sides of the opposition emerge as ideals and practices with their own histories—what we have called mechanical objectivity and trained judgment. (p. 377)

Indeed, this sets the stage for understanding progress in science within a much richer context, in which mechanical objectivity would approximate to the ideals of logical positivism and trained judgment to how science is actually done, namely "science in the making." This clearly shows the importance of understanding the evolving nature of objectivity within a historical perspective.

2.2 Alternative Historical Accounts of Objectivity

According to Fara (2009) during the nineteenth century, Victorian scientists considered Isaac Newton to be the paragon of scientific rationality, "Like a scientific instrument, Newton supposedly recorded neutrally the world about him, and then analysed his data with detachment. Taken to an extreme form, Newton epitomized a pervasive if unattainable scientific stereotype—the selfless genius who measures the Universe as though he were an external observer" (p. 255). For Newton's apple and the law of gravity, see Fara (2015). Interestingly, in most parts of the world school and college science still present this picture of Newton, and it is precisely for this reason that it is important to study the evolving nature of objectivity in a historical context. Next, Fara goes on to critique this eulogy of Newton as a form of objectivity that was questioned among others by the Romantic philosophers during the first half of the nineteenth century. Among these philosophers those following *Naturphilosophen* were outspoken critics as they considered that as human beings we are inextricably entangled with the natural world. In contrast to Newton, Goethe's subjective approach toward scientific experimentation during the eighteenth century included the observer's personal interpretations. For example, as a champion of objectivity, for Newton a prism or a lens is used to produce discrete images that can be inspected with detachment, whereas for Goethe while looking at a prism the retina inside the eye becomes a projection screen (a recording instrument) and thus human beings are inevitably involved in the observations they make. According to Daston and Galison (2007), Goethe and other scientists during the eighteenth century were looking for the idea in the observation and not the raw observation itself, namely an idealization that they considered as *truth-to-nature*.

The next stage in the history of science was characterized by the elimination of all traces of human intervention, idealization, and subjectivity. Both Daston and Galison (2007) and Fara (2009) stress that in order to ensure objectivity it was suggested that human observers be replaced with machines, recording devices, and photographs. According to Fara (2009), Victorian scientists were appalled to think that subjectivity might be at the very heart of science, and similarly Daston and Galison (2007) considered that subjectivity was the enemy within and this led to the implementation of *mechanical objectivity*. Scientists, however, soon found that scientific photographs were equally subject to various experimental factors. This attempt to show the world as it really is faced considerable difficulties, and Fara (2009) presents the dilemma in picturesque terms: "… in order to avoid ending up with a record of the Universe as big as the Universe itself, selections and summaries must be made—an obvious entry point for subjectivity" (p. 258).

Difficulties involved in interpreting the images from recording devices and photographs led the scientists to a new epistemic footprint during the mid-twentieth century, namely "accuracy should not be sacrificed to objectivity" that came to characterize *trained judgment* (Daston & Galison, 2007). In stark contrast, the epistemic regime of mechanical objectivity had endorsed that, "objectivity should not be sacrificed to accuracy." Modern experimental techniques (magnetic maps, x-rays, cloud-chamber photographs) are packed with detailed information which requires the expertise of trained scientists who may differ in their interpretations and hence the confrontation of possible subjective viewpoints (Fara, 2009, p. 259).

Fara (2009, p. 255) considers that the Victorian scientists not only considered Newton to be a paragon of rationality but also made him resemble Nietzsche's "objective man" a passionless being who reflected only that for which he was tuned beforehand. Daston and Galison (2007, p. 250) have endorsed a similar perspective by pointing out that the "objective man" required that the scientist split itself into active experimenter and passive observer. This leads to an essential tension as the objective man of science could be accused of inauthenticity and faced the following dilemma: how could a universally valid working object (e.g., oil drop, cf. Millikan-Ehrenhaft controversy discussed above) be extracted from a particular depicted with all its flaws and accidents? Perhaps the oil drop experiment and the controversy between R. Millikan and F. Ehrenhaft represents this tension and the ensuing dilemma in lucid terms. As suggested by Daston and Galison (2007) in order to go beyond the dilemma Nietzsche smelled the acrid odor of burnt sacrifice. It is plausible to suggest that perhaps Millikan smelled the same odor when he discarded data from almost 59% of the oil drops (Holton, 1978a, b; P. Holton, Email to author, August 3, 2014b; Niaz, 2005, 2015). This leads to the question: What role did objectivity play in Millikan's decision? Fara (2009) provides a possible response in the following terms: "But his [Millikan's] delicate apparatus was easily disturbed, and—armed with his *conviction* that electrons really exist—Millikan discarded around two-thirds of his readings" (p. 325, italics added). To be armed with one's conviction clearly represents a conflicting situation referred to as "self-divided against itself" by Daston and Galison (2007, p. 250).

It is important to note that both Daston and Galison (2007) and Fara (2009) coincide to a fair degree in their presentation of a continuing confrontation between objectivity and subjectivity within a historical perspective. Furthermore, both recognize that the role of human intervention is important in understanding the scientific enterprise and that objectivity has remained an elusive subject. Inclusion of an alternative historical account of objectivity can provide readers a more nuanced understanding of the subject. In this context, it is important to note that Daston and Galison (2007, p. 371) have explicitly endorsed a plurality of visions of knowledge as a permanent aspect of science.

References

Bernard, C. (1865). *Introduction à l'00E9tude de la médecine expérimentale*. Ed. François Dagnognet. (Reprinted, Paris: Garnier-Flammarion, 1966).

Campbell, D. T. (1988a). Can we be scientific in applied social science? In E. S. Overman (Ed.), *Methodology and epistemology for social science* (pp. 315–333). Chicago: University of Chicago Press. (first published in 1984).

Campbell, D. T. (1988b). The experimenting society. In E. S. Overman (Ed.), *Methodology and epistemology for social science* (pp. 290–314). Chicago: University of Chicago Press.

Cushing, J. T. (1989). The justification and selection of scientific theories. *Synthese, 78*, 1–24.

Cushing, J. T. (1996). The causal quantum theory program. In J. T. Cushing, A. Fine & S. Goldstein (Eds.), *Bohmian mechanics and quantum theory: An appraisal* (pp. 1–19). Dordrecht: Kluwer.

Daston, L., & Galison, P. L. (1992). The image of objectivity. *Representations, 40*, 81–128. (special issue: Seeing Science).

Daston, L., & Galison, P. (2007). *Objectivity*. New York: Zone Books.

Dyson, F. W., Eddington, A. S., & Davidson, C. (1920). A determination of the deflection of light by the sun's gravitational field, from observations made at the total eclipse of May 29, 1919. *Royal Society Philosophical Transactions, 220*, 291–333.

Earman, J., & Glymour, C. (1980). Relativity and eclipses: the British eclipse expeditions of 1919 and their predecessors. *Historical Studies in the Physical Sciences, 11*(1), 49–85.

Fara, P. (2009). *Science: a four thousand year history*. Oxford: Oxford University Press.

Fara, P. (2015). That the apple fell and Newton invented the law of gravity, thus removing god from the cosmos. In R.L. Numbers & K. Kampourakis (Eds.), *Newton's apple and other myths about science* (pp. 48–56). Cambridge: Harvard University Press.

Galison, P. (2015a). The journalist the scientist and objectivity. In F. Padovani, A. Richardson & J. Y. Tsou (Eds.), *Objectivity in science*. Dordrecht: Springer. Boston Studies in the Philosophy and History of Science.

Gavroglu, K., & Simões, A. (2012). *Neither physics nor chemistry: a history of quantum chemistry*. Cambridge: Massachusetts Institute of Technology Press.

Gibbs, F. A., & Gibbs, E. L. (1941). *Atlas of electroencephalography*. Cambridge: Cummings.

Gibbs, F. A., & Gibbs, E. L. (1951). *Atlas of electroencephalography: Methodology and controls Vol. 1*. 2nd ed, Reading: Addison-Wesley Press.

Heilbron, J. L. (1981). *Historical studies in the theory of atomic structure*. New York: Arno Press.

Hoffmann, R., Shaik, S., & Hiberty, P. C. (2003). A conversation on VB vs MO theory: A never-ending rivalry? *Accounts of Chemical Research, 36*(10), 750–756.

Holton, G. (1969). Einstein and the 'crucial' experiment. *American Journal of Physics, 37*, 968–982.

Holton, G. (1978a). Subelectrons, presuppositions, and the Millikan-Ehrenhaft dispute. *Historical Studies in the Physical Sciences, 9*, 161–224.

Holton, G. (1978b). *The scientific imagination: case studies.* Cambridge: Cambridge University Press.

Holton, G. (2014a). The neglected mandate: teaching science as part of our culture. *Science & Education, 23*, 1875–1877.

Keats, T. E. (1973). *An atlas of normal Roentgen variants that may stimulate disease.* Chicago: Year Book Medical Publishers.

Millikan, R. A. (1913). On the elementary electrical charge and the Avogadro constant. *Physical Review, 2*, 109–143.

Neuhauss, R. (1898). *Lehrbuch der mikrophotographie* 2nd ed, Brunswick, Germany: Bruhn.

Niaz, M. (1998). From cathode rays to alpha particles to quantum of action: a rational reconstruction of structure of the atom and its implications for chemistry textbooks. *Science Education, 82*, 527–552.

Niaz, M. (2000). The oil drop experiment: a rational reconstruction of the Millikan-Ehrenhaft controversy and its implications for chemistry textbooks. *Journal of Research in Science Teaching, 37*, 480–508.

Niaz, M. (2005). An appraisal of the controversial nature of the oil drop experiment: is closure possible? *British Journal for the Philosophy of Science, 56*, 681–702.

Niaz, M. (2009). *Critical appraisal of physical science as a human enterprise: dynamics of scientific progress.* Dordrecht: Springer.

Niaz, M. (2012). *From 'Science in the Making' to understanding the nature of science: an overview for science educators.* New York: Routledge.

Niaz, M. (2015). That the Millikan oil-drop experiment was simple and straightforward. In R.L. Numbers & K. Kampourakis (Eds.), *Newton's apple and other myths about science* (pp. 157–163). Cambridge: Harvard University Press.

Niaz, M. (2016). *Chemistry education and contributions from history and philosophy of science.* Dordrecht: Springer.

Niaz, M., & Coştu, B. (2009). Presentation of atomic structure in Turkish general chemistry textbooks. *Chemistry Education Research and Practice, 10*, 233–240.

Niaz, M., & Coştu, B. (2013). Analysis of Turkish general chemistry textbooks based on a history and philosophy of science perspective. In M. S. Khine (Ed.), *Critical analysis of science textbooks: evaluating instructional effectiveness* (pp. 199–218). Dordrecht: Springer.

Niaz, M., Kwon, S., Kim, N., & Lee, G. (2013). Do general physics textbooks discuss scientists' ideas about atomic structure? A case in Korea. *Physics Education, 48*(1), 57–64.

Perl, M. L. (2004). The discovery of the Tau Lepton and the changes in elementary-particle physics in forty years. *Physics in Perspective, 6*, 401–427.

Perl, M. L., & Lee, E. R. (1997). The search for elementary particles with fractional electric charge and the philosophy of speculative experiments. *American Journal of Physics, 65*, 698–706.

Ramón y Cajal, S (1989). *Recollections of my life* (trans: Horne, E. & Cano, J.). Cambridge: MIT Press.

Rodríguez, M. A., & Niaz, M. (2004a). The oil drop experiment: An illustration of scientific research methodology and its implications for physics textbooks. *Instructional Science, 32*, 357–386.

Rodríguez, M. A., & Niaz, M. (2004b). A reconstruction of structure of the atom and its implications for general physics textbooks. *Journal of Science Education and Technology, 13*, 409–424.

Stanley, M. (2007). *Practical mystic: religion, science and A.S. Edington.* Chicago: University of Chicago Press.

Wilson, D. (1983). *Rutherford: simple genius.* Cambridge: MIT Press.

Chapter 3
Understanding Objectivity in Research Reported in the Journal *Science & Education* (Springer)

3.1 Method

The journal *Science & Education* (Springer, http://www.springer.com/11191) started publishing in 1992 with Michael R. Matthews (University of New South Wales, Australia) as its Editor. This journal specifically deals with the contributions of history, philosophy, and sociology of science to science education, and is indexed in the Social Sciences Citation Index (Thomson-Reuter). Consequently, it seems that an evaluation of literature published in this journal related to objectivity can help science educators to better understand the evolving nature of objectivity in the history of science. It is interesting to note that Daston and Galison (1992) first presented their ideas with respect to the historical evolution of objectivity (same year that *Science & Education* started publishing), which were later elaborated in Daston and Galison (2007).

In November 2014, I made an online literature search on the website of *Science & Education*, with the keyword "objectivity" (http://www.springer.com/11191). This gave a total of 180 articles published between 1992 and November 2014. All articles were downloaded and a preliminary examination showed that 45 articles could not be included in the study due to the following reasons: (a) Book reviews in which the reviewer refers to the subject of objectivity and not the original author; (b) Book notes, for the same reason as for book reviews; (c) Golden oldies, which included articles by famous historians/philosophers of science written much earlier than 1992; and (d) In some articles the authors provided a reference and the word "objectivity" appeared in the title of that reference.

3.1.1 Grounded Theory

Grounded theory (Glaser & Strauss, 1967) provides a set of guidelines that helps to focus on data collection procedures, based on successive levels of data analysis and conceptual understanding. In the present study, I first classified the selected

© Springer International Publishing AG 2018   37
M. Niaz, *Evolving Nature of Objectivity in the History of Science and its Implications for Science Education*, Contemporary Trends and Issues in Science Education,
DOI 10.1007/978-3-319-67726-2_3

articles from *Science & Education* in different levels (details are presented below), which were later assigned a category, and finally in Chap. 7, categories from different studies (Chaps. 3–6) are compared to facilitate conceptual understanding. This procedure can be summarized in the following steps: (a) Comparison of data sources (articles) to assign a level (I–V); (b) comparison of these levels (presented later) which facilitated their classification in categories; and (c) comparison of categories from different studies to facilitate understanding and draw conclusions. Following guidelines were used while developing the different steps of the procedure (based in part on Charmaz, 2005, p. 528):

1. Familiarity with the setting and topic of study in each of the selected articles.
2. Evaluate classification of the selected articles to see if they are based on appropriate evidence.
3. Systematic comparisons between the classifications and the categories.
4. The need for the categories to represent a wide range of experiences represented in the classifications.
5. Establish a logical and conceptual link between the classifications, categories, and arguments for the analyses.

Although the guidelines presented above were of considerable help in different stages of data analysis, a word of caution is necessary: "… grounded theory does not refer to some special order of theorizing per se. Rather, it seeks to capture some general principles of analysis, describing *heuristic strategies* that apply to any social inquiry independent of the particular kinds of data: indeed it applies to the exploratory analysis of quantitative data as much as it does to qualitative inquiry" (Atkinson & Delamont, 2005, p. 833, italics added). The emphasis on heuristic strategies is particularly important in the present study, as they facilitated conceptual understanding.

3.1.2 Classification of Articles

Finally, a total of 131 articles were evaluated and classified in the following levels (criteria for evaluation are based primarily on Daston & Galison, 2007):

Level I Traditional understanding of objectivity as presented in science textbooks and some positivist philosophers of science. It is based on an ideal of objectivity as an important human value and part of the scientific outlook.

Level II A simple mention of objectivity as an academic/literary objective. It recognizes that although science is not value free, but still this does not affect the objective status of science.

Level III The problematic nature of objectivity is recognized. However, no mention is made of the changing/evolving nature of objectivity.

Level IV An approximation to the evolving/changing nature of objectivity, based on the social and cultural aspects of objectivity.

Level V A detailed historical reconstruction of the evolving nature of objectivity in the history of science that recognizes the role of the scientific community and its implications for science education.

Following the guidelines presented above (cf. Charmaz, 2005), and in order to facilitate credibility, transferability, dependability, and confirmability of the results I adopted the following procedure: (a) All the 131 articles from *Science & Education* were evaluated and classified in one of the five levels; (b) After a period of approximately three months all the articles were evaluated again and there was an agreement of 90% between the first and the second evaluation; and (c) After another period of three months all the articles were evaluated again, and there was an agreement of 92% between the second and the third evaluation. This procedure was particularly helpful in understanding the underlying issues as according to Denzin and Lincoln (2005): "Terms such as *credibility, transferability, dependability*, and *confirmability* replace the usual positivist criteria of internal and external validity, reliability, and objectivity" (p. 24, original italics).

A complete list of all the 131 articles from *Science & Education* that were evaluated is presented in Appendix 1. In the section on Results and Discussion, 71 examples of the different levels are provided, with the following distribution: Level I = 2, Level II = 15, Level III = 42, Level IV = 10, and Level V = 2. These examples provide an understanding of how the subject of objectivity has been discussed by authors in this journal. It is important to note that all the articles evaluated in this study referred to objectivity in some context, which may not have been the primary or major subject dealt with by the authors. Detailed examples of all five levels are presented in the next section. Distribution of all the articles according to author's area of research, context of the study, and level (classification) is presented in Appendix 2.

3.2 Results and Discussion

Each of the 131 articles from *Science & Education* was evaluated (Levels I–V) with respect to the context in which they referred to objectivity. Based on the treatment of the subject by the authors following 37 categories (sections) were developed to report and discuss the results (cf. guidelines presented above from Charmaz, 2005). These categories along with the examples are presented in alphabetical order. It is important to note that some of the articles could easily be placed in more than one category. The idea behind the creation of 35 categories (sections) is to facilitate the reader to find the subject of her/his interest. It is important to note that *Science & Education* has a readership and contributors that include science educators, historians, philosophers of science and sociologists that cover many areas of the science curriculum. Given the wide range of subjects discussed by the authors over a period of more than 20 years, it is difficult to create the semblance of a continuous storyline (as suggested by one of the reviewers).

For example, in the 1990s constructivism was a subject of considerable importance, and in recent years the research community seems to have lost interest in it.
Similarly, due to limitations of space it is not possible to present a detailed critical
analysis of every article. Complete information about each article and the author is
provided in the appendices (1 and 2) which can be consulted by the interested
readers. Next, examples from the 35 categories are presented.

3.2.1 Argumentation and Objectivity

The role of argumentation in the classroom has been the subject of considerable
research in the science education literature. Drawing on the work of Longino
(1990, 2002), Jiménez-Aleixandre (2012) has explored the relationship between
objectivity in science and explanatory plurality:

> Longino (1990) undertook an analysis of scientific knowledge with the goal of reconciling
> the objectivity of science with its social and cultural construction. Recently she has
> explored the epistemological consequences of the recognition of the social character of
> scientific inquiry in connection to pluralism, or the acknowledgement of *explanatory plur-
> ality* (Longino, 2002). For Longino (2008) knowledge itself is social, because what mat-
> ters is what the scientific community comes to agree or disagree on …. Viewing scientific
> knowledge as socially constructed has influenced both the design of science classrooms as
> communities of learners, and the ways of studying classroom interactions, in particular
> the discursive ones, as argumentation (p. 469, italics added). Classified as Level IV.

With this background the author has followed argumentation in genetics classrooms requiring models to build explanations, which leads to the framing of genetics issues in their social context. Campbell (1988a) a methodologist had referred
to "explanatory plurality" as plausible rival hypotheses, quite similar to Longino.
The presentation of Jiménez-Aleixandre (2014) comes quite close to what Daston
and Galison (2007) have referred to as trained judgment.

3.2.2 Classification of Species and Objectivity

According to Takacs and Ruse (2013), classification presents a number of interesting issues in the philosophy of biology:

> Everybody recognizes that there is a certain degree of subjectivity involved in classifica-
> tion, so much so that there is sometimes debate about whether classification is a science
> or an art. However, it is generally agreed that at the lowest level, the level of species, there
> is significantly more reality or objectivity. No one, for instance, thinks that it is a matter
> of choice about whether Michael Ruse or Peter Takacs should be included in the group
> Homo sapiens, and that Toto the dog and Secretariat the horse should be excluded. The
> question now becomes that of wherein lies the objectivity or reality of species, as opposed
> say to genera (p. 23). Classified as Level IV.

Authors also go beyond by pointing out the subjectivity involved in for example in the inclusion of *Homo sapiens* along with *Homo erectus* and *Homo habilis*

in the genus *Homo*. Again they raise the issue of whether there would be consensus in including the *Australopithecus afarensis* (to which the famous fossil Lucy belongs) in the genus *Homo*. This clearly shows how different interpretations lead to controversies that produce tension in our understanding of the objectivity–subjectivity duality.

3.2.3 Commodification of Science and Objectivity

Commodification and commercialization of science has been the subject of recent research in science education (see the special issue edited by G. Irzik, 2013). This research shows that scientific knowledge becomes more and more like a commodity as part of the market economy in which the influence of money and corporate research become dominant. In some cases universities and research institutions become increasingly organized like a private company.

In this context, according to Vermeir (2013):

> These *basic characteristics and norms of science* may be lost with increasing commodification. Current science policy sees some of the positive and constitutive properties of science as obstacles, because they hinder the commodification and market adaptation of science. Legislation and policy try to remedy these perceived "obstacles" by social engineering: the nonexcludability, positive externalities and cumulativeness of scientific knowledge are reduced by intellectual property regimes, for instance; the importance of trust and values are replaced by standardization; expert judgment and peer-to-peer self-regulation are replaced by techniques of mechanical objectivity (pp. 2506–2507, italics added, footnote states: "For mechanical objectivity and expert judgment as different regimes of objectivity, see Daston & Galison, 2007"). Classified as Level IV.

The *basic characteristics and norms of science* refer to the Mertonian norms that include: sharing and openness in scientific practice, truthfulness, objectivity, trust, accuracy, and respect for expertise (Merton, 1979). The transition from trained judgment to mechanical objectivity in the context of commercialization of science is a cause of concern for Vermeir and perhaps also for many science educators. However, according to Daston and Galison, the transition from one extreme (mechanical objectivity) to another (trained judgment) can go back and forth.

3.2.4 Consciousness and Objectivity

According to Marroum (2004):

> What complicates the objectivity of any educational study is that unlike scientific research, which deals with sensible data, educational research must also deal with the data of consciousness (of both students and teachers). Teachers' perceptions and beliefs about learning significantly affect how they approach the material and what they teach. The same can be said of students, and their perceptions affect how they learn. Teachers who follow the inquiry approach to teaching, for example, have varying conceptions of what

inquiry means. Thus, a theory adopted by different teachers can lead to contrary results. This might provide a clue as to why some research shows that teaching standard textbook physics does not produce significant changes in the conceptual understanding of the material, while others show the contrary (pp. 538–539). Classified as Level II.

Marroum's work is based on the cognitional theory of Bernard Lonegran, who does not provide ready-made answers to readers. His approach requires teachers to first self-appropriate what they are teaching to the students. It facilitates the integration of the history of science into the curriculum. He suggests that when students discover what they have in common with Archimedes, Aristotle, Galileo, Newton, Maxwell, and other scientists, they will develop confidence in their ability to learn (It is not clear if Marroum follows this historical approach. For further details on Lonegran's theory, see Roscoe, 2004). Furthermore, in order for learning to be meaningful, the student must move beyond subjective knowledge to objective knowledge.

3.2.5   *Constructivism and Objectivity*

Given the considerable amount of controversy in the science education literature with respect to radical and social constructivism, this section has the following four presentations: Suchting (1992), Slezak (1994), and Garrison (1997, 2000). However, in recent years interest in constructivism has declined.

In the context of his criticism of the subjective realism espoused by radical constructivism (Ernst von Glasersfeld), Suchting (1992) clarifies that contrary to popular belief, immutability and certainty have nothing essential to do with our understanding of objectivity (p. 226). For example, the Galilean transformation equations of classical kinematics proved not to be immutable, as they are replaced in special relativity by different and more general equations. Similarly, the approximations in Galilean equations are not less objective than the previously non-approximate ones. The other characteristic that sometimes is invoked to understand objectivity is certainty. For example, the statement that "Isaac Newton was born on 4 January 1643" is considered to be certain and an instance of objective knowledge. However, even such statements are problematic as the information included may be erroneous or false. In this context, for Suchting (1992), understanding of immutability and certainty show the problematic nature of objectivity. Classified as Level III.

According to Slezak (1994):

> Besides the facts and theories conveyed in a science education are certain values and norms of conduct. Some of these are more specifically pertinent to the practice of science, while others are general moral precepts of the community at large. Besides the academic conventions concerning citations, acknowledgments and other scholarly practices are the noble ideals of objectivity and truth which have been seen as among the important human values embodied in the scientific outlook. The inculcation of these broader values has been widely taken to be among the important functions of a science education, but the doctrines of social constructivism may be seen as posing a fundamental challenge to this ethical dimension of science education as well (p. 269). Classified as Level I.

In order to facilitate the ethical dimensions of science (which may be weakened by social constructivism) the author endorses Merton's "ethos of science" (p. 270). Furthermore, the traditional conventions regarding scientific publications have been the subject of considerable controversy in the history and philosophy of science literature (e.g., Medawar, Holton, Polanyi), as they depart from how science is actually done, namely "science in the making" (cf. Niaz, 2012).

Garrison (1997) critiques Von Glasersfeld's radical constructivism as subjectivist and instead recommends Deweyan social constructivism based on experimentalism as an alternative:

> The difference between subjectivist constructivism and social constructivism comes down to the difference between practical overt operations of inquiry (for example, experimental science), and the occult internal operations of "mind" characterized by von Glasersfeld's "mental operations" at the level of *reflective abstraction.* For the pragmatist a clean shave with Ockham's razor whisks away von Glasersfeld's needless subjectivism and mentalistic abstractions, thereby clearing the face of reasonable science education for genuine experimentalist and objective social constructivism (p. 553, original italics). Classified as Level II.

Garrison (2000) also refers to Ernst von Glasersfeld's constructivism as subjectivist: "It is a peculiarly subjectivist form of constructivism that should not be attractive to science and mathematics education concerned with retaining some sort of realism that leaves room for objectivity" (p. 615). Garrison ignores the historical context in which objectivity is always achieved in degrees, namely the recognition that it is a process. It is plausible to suggest that Garrison's position approximates to an academic form of objectivity that is Level II. In the framework of Daston and Galison (2007), both presentations by Garrison (1997, 2000) represent mechanical objectivity.

3.2.6 Controversy and Objectivity

According to Hildebrand, Bilica, and Capps (2008), controversies in science education are more intractable than those in science as they involve a wider range of considerations, such as epistemic, social, ethical, political, and religious. Authors then consider the controversy between Intelligent Design Creationism (IDC) and evolution and present the following possible strategies generally used in the biology classroom: (a) Teach the controversy—this strategy assumes that students should be allowed to make up their own minds on controversial issues; (b) Avoidance—in this case teachers may choose to omit controversial topics; and (c) Dogmatism—this alternative would dismiss the controversy altogether. In contrast, these authors suggest a proactive, philosophically pragmatic approach based on the work of John Dewey (1925/1983), according to which knowledge is achieved primarily through a process of inquiry that is characterized by its social, experimental, and fallible nature. Furthermore, inquiry begins for most people not with abstract puzzles but with concrete problematic situations. This approach neither avoids nor ignores controversy and thus goes beyond the *narrow epistemological solutions* generally presented in school science:

> In consequence, this means that narrow epistemological solutions will often be insufficient to resolve controversies in science education: it cannot be enough to prove a particular theory is "true" or "verified." Consider an example that illustrates this: the proponents of IDC advocate for a "teach the controversy" approach to teaching evolution. This pedagogical approach, proponents argue, is necessary because of the scientific community's commitment to "objectivity" and "fairness." To exclude some views would amount to the unfair marginalization of an unpopular view (Hildebrand et al., 2008, p. 1036). Classified as Level III.

The problematic nature of objectivity in this presentation is quite peculiar. Proponents of IDC support a commitment to objectivity as this would allow them to include their ideas with respect to evolution. This clearly shows how biology teachers may have to be more thoughtful while introducing objectivity in the classroom.

Following a historical reconstruction of the topic of chemical equilibrium in the chemistry curriculum, Quílez (2009) has suggested that the inclusion of such details can motivate students to study chemistry and even perhaps understand the underlying controversial ideas. According to the author: "Objectivity, certainty and infallibility as universal values of science may be challenged studying the controversial scientific ideas in their original context of inquiry ..." (p. 1204). Classified as Level III. This seems to be sound advice for making the science curriculum more relevant for the students.

3.2.7   Discovery and Objectivity

According to Kipnis (2007), learning about discovery helps students to understand how scientists work. This led him to conclude that discovery is objective in the sense that having been created it exists forever and cannot be undone: "As to the discovery, if it is done, it is done; it acquires a certain objectivity which no subsequent labeling can remove" (p. 907). This presentation ignores the social context in which scientific discoveries are evaluated, critiqued, accepted, reinterpreted, and eventually even changed by the scientific community. Classified as Level II.

3.2.8   Disinterestedness and Objectivity

Kolstø (2008) has argued that the post-academic science differs from academic science in the past, and the inclusion of history of science in the curriculum can facilitate democratic participation and the disinterested pursuit of objective truth. Finally, the author concluded: "Furthermore, in the post-academic mode of research, the scientists' autonomy is reduced. Although the researchers might have autonomy on the more detailed level, the problem area to be studied is typically defined by the funding agency. Thus, the typical post-academic scientist has become a contractor and has to make dispositions that might give him research contracts. Such research funding relationships makes it hard to claim full objectivity and disinterestedness" (p. 980). Classified as Level II. Achieving "full objectivity" is a complex *process* and needs to go beyond being disinterested.

3.2.9   Diversity/Plurality in Science and Objectivity

Allchin (2004) has explored the history of craniology and phrenology to show that these were considered to be scientific endeavors, based on huge amounts of data, considered as a "Baconian orgy of quantification" in the nineteenth century. For several decades anthropologists, such as Paul Broca, tried to use skull measurements to prove sexual and racial differences in intelligence. At the time, however, craniology seemed like a straightforward application of the principle of structure and function, namely if mental functions take place in the brain, then the brain's size should reflect mental capacity. Similarly, phrenology, the study of cranial shapes and proportions seemed very plausible:

> Moreover, craniology was quantitative, following one oft-cited hallmark of science. Craniologists used over 600 instruments and 5,000 measurements … Of course, the prospects of craniology and phrenology went unfulfilled. When women eventually entered the field, they challenged claims earlier deemed acceptable by men. Standards of evidence rose. The whole field soon dissolved. In retrospect one can see that the community of (white) European male researchers was culturally biased (not that any practitioner recognized his own bias). Now the episode is a persuasive example of *how diversity in a scientific discipline can contribute to its objectivity* …. Craniology is wrong, not misguided. History thus offers complementary lessons in science and pseudoscience. It helps reveal vividly how science works and why, sometimes, it errs (Allchin, 2004, pp. 190–191, italics added). Classified as Level III.

The reference to "Baconian orgy of quantification," instruments and measurements in the nineteenth century approximates to Daston and Galison's (2007) mechanical objectivity. However, Allchin's perspective does not foresee the transition from mechanical objectivity to trained judgment, but rather emphasizes that *diversity in a scientific discipline can contribute to its objectivity*. In a sense this approximates to the interpretation of science as social knowledge as suggested by Longino (1990).

Carrier (2013) has outlined the role played by values, value-ladenness, and pluralism in understanding objectivity in scientific development based on the following facets of history of science: (a) The traditional notion of objectivity was strongly shaped by Francis Bacon (p. 2549). Bacon's notion of objectivity required the scientist to be neutral and detached from the research project; (b) Contrary to Bacon's rules, history of science shows that values play an important role in the development of science as facts/data in and by themselves do not determine how they are to be interpreted; (c) Values tend to be contentious and thus can be regarded as a threat to scientific objectivity; (d) As Baconian objectivity is hard to follow, pluralism based on value-judgments is a virtue rather than a liability; (e) The social notion of objectivity was introduced by Popper (1962) and Lakatos (1970) and focuses on conflicting approaches adopted by scientists; (f) Longino (1990) has recommended science as social knowledge as the pluralist approach to objectivity helps to correct flaws and thus enhance the reliability of scientific results. Longino is widely considered to have undermined or dissolved the distinction between the epistemic and the social; (g) Pluralism remains as a step in the development of science and eventually gives way to consensus. This is supported by Kuhn's normal science and also based on the work of Kitcher (1993), Laudan (1984) and Collins and Evans (2002). Finally, Carrier (2013) concluded that pluralism does not detract from scientific objectivity but is a means to achieving objectivity: "Scientific consensus formation is possible because, regardless of divergent epistemic inclinations and predilections, scientists have a fundamental commitment in common, the commitment, namely, to give heed to certain rules in debating knowledge claims. Adopting such rules serves to curb subjective preferences for the sake of producing knowledge that enjoys intersubjective assent" (p. 2565). Classified as Level V. An important aspect of this presentation is the emphasis on a pluralistic value-laden nature of scientific judgments, within a historical context that facilitates an intersubjective consensus in the scientific community.

3.2.10 Enrollment Practice and Objectivity

In the 1960s the Swedish government became concerned of the declining number of students who chose to study science as a career. Based on this in the 1970s and 1980s, initiatives were taken to make science more attractive and a fun subject to students, referred to as the TEK-NA projektet (1975). This campaign to foster interest in science led to a conflict as some sectors of the society perceived it as a threat to an individual's right to a free choice. Lövheim (2014) depicts the dilemma in the following terms:

> The TEK-NA project also targeted student counselors in their strategy to achieve a change of attitudes. This confirmed the belief in career guidance as a way of creating positive propaganda; the Swedish government had stressed the need for such a development during the 1970s .... Consequently student counselors were involved as a direct channel to pupils approaches to science. As a technology of government they were part of every-day school life without interfering with direct class room practice .... The text also contained sections with advices on how to guide pupils—especially girls—into identities as engineers or scientists ... the project lead to protests from student counselors who claimed they were forced to persuade pupils into the high school Science program and that the material lacked a sense of objectivity ... (pp. 1776–1777). Classified as Level II.

This is an interesting example of how some reform efforts (more experiments and less abstract textbooks) can be construed to be less rigorous than the traditional science curriculum and thus lack objectivity. Similar relationship between traditional science and objectivity can also be found in other countries.

3.2.11 Evolution, Creationism and Objectivity

Difficulties involved with these complex and controversial subjects is referred to by Smith, Siegel, and McInerney (1995) in the following terms: "It is important to note, however, that good science seeks to be as objective and impartial as possible. The expert scientist not only recognizes that his work may be influenced by personal biases but also overtly seeks to identify and eliminate improper influences" (p. 29). Classified as Level III.

With respect to teaching creationism in public schools, Pennock (2002) stated:

> The charge that such a policy violates academic freedom is not so easily dismissed. One might reasonably dispute about whether academic freedom applies in the public elementary and secondary schools in the same way that it does in higher education, but primafacie there seems to be no good reason to think that this important protection should be afforded to university professors and not to others of the teaching profession who serve in other educational settings. However, academic freedom is not a license to teach whatever one wants. Along with that professional freedom comes special professional responsibilities, especially of objectivity and intellectual honesty. Neither "creation-science" nor "intelligent-design" (nor any of the latest euphemisms) is an actual or viable competitor in the scientific field, and it would be irresponsible and intellectually dishonest to teach them as though they were (Pennock, 2002, p. 121). Classified as Level II.

Finally Pennock concluded that neither "creation-science" nor "intelligent-design" is an actual or viable competitor in the scientific field, and based on objectivity it would be irresponsible and intellectually dishonest to teach them as though they were. Although this may seem to be sound advice, at least some science educators may not agree with it.

Homchick (2010) has studied the controversy between the evolutionists and the creationists in the context of the American Museum of Natural History's Hall of the Age of Man during the early 1900s. Henry Fairfield Osborn, president of the museum based his curatorial work on the purported use of objectivity as a means to communicate the validity of the evolutionary theory. However, this was criticized by the Baptist pastor John Roach Straton by establishing a different type of objectivity based on pluralistic approaches to theories of origin that included both evolutionary theory and creationist account. Consequently, established as a common value, objectivity ceased to discriminate between scientists and nonscientists. Next, Homchick considers that both Daston and Galison (1992) and Gergen (1994) provide useful lenses to look at the Osborn-Straton debate. With respect to the historical origin of objectivity, Homchick (2010, p. 486) noted:

> Objectivity, often connected with the rise of Baconian science, came to be associated with a particular matrix of values in the nineteenth century. Lorraine Daston and Peter Galison in their article, "The Image of Objectivity," discuss the use of objectivity during and after the nineteenth century (Daston & Galison, 1992). They identify atlases as bearers of the concept of objectivity specifically because of the association between the visual and the factual embedded in this type of artifact. Additionally, the authors establish how objectivity is not only powerful through the visual content, but that the use of this concept actually represented an apparent superiority of judgment through a "self-denying moralism." (Daston & Galison, 1992, p. 99)

Similarly, according to Homchick, Gergen (1994) considers objectivity not to be a static characteristic of texts and objects and differentiates objectivity through two general categories that of process and product. Thus, it seems that Osborn relied primarily on the objectivity of the product, namely the artifacts displayed in the museum exhibit. In contrast, Straton used the objectivity of process to criticize Osborn for not including the creationist account. Finally, Homchick (2010) concluded:

> Here Osborn appears to embody Daston and Galison's identification of objectivity as allowing "nature to speak for itself" (Daston & Galison, 1992, p. 81) and Gergen's identification of objectivity as surfacing through the "true" character of the natural world. In this formulation, objectivity emerges through the product—the artifact of nature (p. 491). Classified as Level V.

Daston and Galison (2007) refer to this form of objectivity as "truth-to-nature." The Osborn-Straton controversy also shows how the pluralistic approach to science (Giere, 2006a, b) can also be used not only for promoting the scientific endeavor but also the creationist account. Such controversies can provide teachers an opportunity to include topics in the classroom that can lead to lively discussions.

3.2.12   Expert Knowledge and Objectivity

Lindahl (2010) has investigated students' reasoning about conflicting values concerning the human–animal relationship exemplified by the use of genetically modified pigs as organ donors for xenotransplantation:

> The students' use of scientific knowledge (expert knowledge) as well as personal or everyday knowledge (embedded in local practice) in arguments was used to deepen the analysis of the students' understanding and to discern their appreciation of expert knowledge and disembedded practices. The use of scientific knowledge for their argumentation was regarded as an appreciation of expert knowledge, and their support for biotechnology relating to the discussed example was interpreted as their appreciation of disembedded practices. Typically, the use of expert knowledge was seen as a way to create objectivity and distance to the dilemma …. When a student contradicted his/her contextualized argument with expert knowledge, it was seen as an attempt to objectify (p. 885). Classified as Level II

Following is an example of an episode in which expert knowledge was manipulated by a government for its own political agenda. According to Legates et al. (2015):

> A better approach to determining an appropriate methodology to identify and quantify a consensus can be found in the work of Lefsrud and Meyer (2012). They argue that building a consensus "fundamentally depends upon expertise, ensconced in professional opinion" (p. 1478). Even here, a Classical purist might legitimately argue that appealing to the authority of experts, however well qualified, is the Aristotelian logical fallacy later labeled by the medieval schoolmen as the argumentum ad verecundiam—the argument from reputation. Experts can be unanimously wrong, as the case of the 100 German

authors who opposed Einstein's theory of relativity in the years leading to World War II. They were wrong because the regime demanded them to make scientific objectivity subservient to the racial politics of the regime (p. 12). Classified as Level III.

This episode provides an interesting and thought-provoking backdrop to Daston and Galison's (2007) regime of trained judgment as an alternative to mechanical objectivity based on expert knowledge. In other words the opinion of the experts can be politically motivated and hence the difficulties involved in accepting trained judgment as an alternative to mechanical objectivity.

Allagaier (2010) has explored the role of scientific experts in the creation/evolution controversy as presented in the UK press:

Following traditional accounts of expertise, a scientific expert is a formally trained specialist in a scientific discipline …. The scientific community developed through professionalisation and formal training and established a professional ideology … in which they portray themselves as value-free, neutral and objective experts …. However, from a sociological point of view, scientists cannot operate outside society; they are as much members of the public as anyone else. The notion that a scientific expert can be entirely neutral, value-free and objective cannot be sustained from a sociological perspective (e.g., Restivo, 1994). (p. 800). Classified as Level III.

The presentations by Allagaier (2010) and Legates et al. (2015) provide interesting examples with respect to the role played by experts and expert knowledge in modern society. As part of society experts also have difficulty in being entirely objective and value-free. Perhaps similar constraints can also be observed in the peer-review process used by most scientific journals.

3.2.13 Feminist Epistemology and Objectivity

Based on a critical appraisal of feminist epistemology (Harding, Keller, & Pinnick), Ginev (2008) has advocated a theory of gender plurality that leads to a conception of dynamic objectivity. Harding (1987) considers that using women's lives as grounds to criticize the dominant forms of scientific knowledge can decrease the partialities in the picture of the world presented by the natural sciences. Keller (1985) has suggested a multi-gendered scientific research that leads to the idea of dynamic objectivity. Pinnick (2005) is, however, more critical by asserting that there are no data that would test the validity of the hypothesis that there is a causal relationship between women's lives and science's cognitive ends.

Finally, Ginev (2008) concluded: "In a hierarchically organized society, objectivity cannot be defined as requiring value-neutrality: The politically engaged standpoint of feminism is less partial and distorted than the standpoint of conventional scientific inquiry. By implication, the former should lead to pictures of nature and social relations that are 'more objective' than those obtained by means of the existing natural and social sciences" (p. 1142). Classified as Level III. This shows that we need to explore the degree to which a field of inquiry has achieved objectivity.

3.2.14   Genetics, Ethics and Objectivity

Blake (1994) has analyzed three pioneer programs (at three universities in USA) that attempt to integrate genetics and ethics in the classroom. A major critique of the study is the lack of continuity between the pedagogical goals and the theoretical framework of these programs. The programs adhered to an underlying framework based on "tacit assumptions" (Keller, 1992, p. 27) that undercut the veracity of ethics, and emphasized reason, empirical evidence, and objectivity. Finally, Blake (1994) concluded:

> The curricular possibilities of the "new genetics" for the science classroom—gel electrophoresis of DNA fragments, recombination of DNA into bacterial plasmids—have a similar intoxicating effect which distracts the science educator from the task of critical reflection on the "tacit assumptions" of their programs. This is not merely a priority of science over ethics in the science classroom but a much more fundamental disparity. This modern view of science and consequent epistemological privilege have been critically examined by philosophers, sociologists and historians of this century (cf. Feyerabend, 1975; Keller, 1992; Kuhn, 1962; Lakatos, 1970; Midgley, 1985) .... The ideals of objectivity, rationality and empirical privilege have been seriously and soundly challenged .... Science has an historical and social context; science is contingent and subjective (p. 387). Classified as Level III.

This presentation was classified as Level III as it clearly shows the problematic nature of objectivity. Furthermore, Blake (1994) refers to two major issues that are of considerable importance to science education. First, she refers to the problem of two cultures, introduced by C.P. Snow (1963), namely a gulf of mutual incomprehension between the literary intellectuals and the scientists. Second, based on Keller (1992) she asserts that scientists are probably less reflective of "tacit assumptions" that guide their reasoning than any other intellectual of the modern age. Indeed, this is all the more ironic as Polanyi's (1966) *tacit dimension* was published almost half a century ago. Polanyi (1964, 1966) differentiated between two kinds of knowledge: (a) explicit, articulated, and formal knowledge; and (b) tacit, unarticulated, and non-formalized knowledge. He argued that the first cannot be achieved without the second. These considerations led Polanyi to question the false ideal of "objectivity" in post-Enlightenment scientific thinking.

3.2.15   Historical Contingency and Objectivity

The contingent nature of science has been recognized by physicist-philosopher James Cushing (1989). According to Cushing (1995), David Bohm's (1952) work can be seen as an exercise in logic, thus providing evidence that the Copenhagen interpretation of quantum mechanics was not the only logical possibility compatible with the facts:

> Given the presumed objectivity and impartiality of the scientific enterprise, one might expect that such an interpretation [Bohm's] would have been accorded serious consideration by the community of theoretical physicists. However, it was basically ignored, rather

than either studied or rebutted. Just as external factors had played a key role in establishing the Copenhagen hegemony, so they once again contributed to keeping this competitor from the field. That a generation of physicists had been educated in the Copenhagen dogma made it all the more difficult for Bohm's theory (Cushing, 1995, pp. 139–140). Classified as Level III.

According to the contingency thesis, the same experimental observations can be explained by rival theories (in this case the Copenhagen and Bohm's interpretation of quantum mechanics). In other words the order in which events take place is an important factor in determining which of two observationally equivalent theories is accepted by the scientific community. With respect to the presumed objectivity of the scientific enterprise, it is interesting to note that Bell (1987) a leading scholar on the Bohmian interpretation of quantum mechanics has raised the following thought-provoking questions: (a) Why is the pilot wave picture (de Broglie and Bohm's ideas) ignored in textbooks; and (b) Should Bohm's interpretation of quantum mechanics not be taught?

At this stage it would be interesting to consider a possible relationship between Cushing's idea of *contingency* and the historical evolution of the regime of objectivity as presented by Daston and Galison (2007). In other words, it is plausible to suggest that it is perhaps the contingent nature of science (among other factors) that manifests itself in the evolving nature of objectivity. Furthermore, it can be argued that the Copenhagen and the Bohm interpretations of quantum mechanics constitute an example of methodological pluralism in the history of science.

3.2.16 Historical Narratives and Objectivity

Kubli (2007) has emphasized the need to go beyond the simple regurgitation of experimental details, and provide students with the historical narratives (stories) which provide the background to understanding progress in science:

> Of course, scientific reasoning and laws can be imparted in a completely objective way: they can be reduced to facts and figures without any human element, and indeed, some scientists and even teachers see such *objectivity as the characteristic of true science.* Of course, scientific laws are independent of the specific circumstances of their discovery. They can be "proved" by a reproduction of the basic experiments—which can be repeated whenever there is a need to do so .... This approach has not disappeared, even among teachers, in spite of engaged discussions in science education. It stands in contrast to the view that, in science teaching, stories are not only justified, but necessary (Kubli, 2007, p. 519, italics added). Classified as Level III.

This presentation shows the need to go beyond the traditional forms of objectivity (and hence its problematic nature) by incorporating the human element involved in scientific progress in the form of science narratives (stories), especially during "science in the making." According to Klassen (2006): "School science lacks the vitality of investigation, discovery, and creative invention that often accompanies *science-in-the-making* ..." (p. 48, italics added).

3.2.17   *History and Objectivity*

According to Matthews (1992):

> We know that *objectivity in history is, at one level, impossible*: history does not just present itself to the eye of the beholder; it has to be manufactured. Materials and sources have to be selected; questions have to be framed; decisions about the relevant contributions of internal and external factors in scientific change have to be made. All of these matters are going to be influenced by the social, national, psychological, and religious views of the historian. More importantly they are going to be influenced by the theory of science, or the philosophy of science, held by the historian. Just as a *scientist's theory* affects how they see, select, and work upon their material, so also will a *historian's theory* affect how they see, select, and work upon their material (p. 19, italics added). Classified as Level IV.

Interestingly, in the very first issue of *Science & Education*, Michael Matthews as founder Editor has set the tone for what he expected the journal to promote, espouse, and cultivate. At the end of the citation, Matthews provides the well-known quote from Lakatos (1971), to the effect that if philosophy of science without history of science is empty, then history of science without philosophy of science is blind. Rest of the citation constitutes a preamble and even perhaps a guide to future research on the application of history and philosophy of science (HPS) to science education. It refers to the difficulties involved in recounting any historical episode, and hence the problematic nature of objectivity. Interestingly, he draws a parallel between the *scientist's theory* and a *historian's theory*, as both are theory-laden. It is not farfetched to suggest that in the case of a conflict between the two theories, it is the historian's responsibility to set the record straight. A good example of this conflict is the role played by Holton (1978a, b) in the oil drop experiment that helped to understand Millikan's handling of his published data. Matthews (1992) provides another facet of this conflict by referring to the case of Galileo, who was considered by nineteenth-century philosophers and scientists as an inductivist and empiricist. However, this picture changed in the twentieth century and Galileo came to be considered as a Platonist dedicated to rationalism and thought experiments.

3.2.18   *History of Science and Objectivity*

According to Leite (2002):

> Throughout the previous section a few arguments were already put forward to support the idea that the history of science can help students to acquire an adequate image of science. Enabling students to realise that models in science have been altered and modified in order to fit new data and that the same phenomena can be explained by different models, history of science gives students the opportunity to see how scientific knowledge is provisional and uncertain and how, even in science, we cannot find objectivity and truth … (p. 337). Classified as Level III.

Due to the changing nature of scientific models, this presentation emphasizes the tentative nature of scientific knowledge. Leite then goes beyond by associating uncertainty in science with difficulties involved in finding objectivity and truth. The essence of the idea expressed in this presentation is quite similar to what Matthews (1992) had referred to previously with respect to objectivity in history.

Lyons (2010) has stressed that we need to do a better job of teaching students about the process of science. The practice of science is not quite the straightforward objective process that many scientists suggest:

> The history of science documents that determining *what is a "fact" is continually reevaluated in light of ongoing investigations* .... More important, a variety of factors contribute to whether a particular idea is readily accepted, from the prestige of the person advocating it to how well it fits in with prevailing social views .... Nevertheless, objectivity is a value that all scientists strive for in their work. Science is as successful as it is because it has developed a set of standards and a methodology for designing experiments, interpreting results, and constructing effective scientific institutions. This does not prevent scientists from making mistakes, but the various aspects of scientific practice mean that science has enormous capacity to be self-correcting (p. 457, italics added). Classified as Level III.

This presentation attempts to establish a balance between how scientists strive to be objective and that the practice of science shows how various factors are influential in the acceptance of a theory and this often leads the scientists to make mistakes. Science teachers and textbooks generally emphasize that the scientific enterprise is based on "facts." However, this is more complex than it seems at first sight and Lyons rightly points out that, "what is a fact is continually reevaluated."

3.2.19 Marxism and Objectivity

According to Deng, Chai, Tsai, and Lin (2014):

> ... Marxism puts less emphasis on the social/cultural influence on science while highlighting the objectivity and rationality of science (Wan et al., 2013). Another possible explanation can be that school science teaching practice pays relatively less attention to the role of society in science. In China, Marxism tends to highlight relatively more the pragmatic values of scientific knowledge than the influence of society on the development of scientific knowledge (p. 853). Classified as Level II.

At first sight, this may appear somewhat counter-intuitive, given the strong relationship between Marxism and changes in society. However, the authors go on to clarify that based on the work of Mao (1986), the concept of "practice" has been emphasized and consequently highly valued in China. Mao even considers practice as the sole criterion for testing truth and value of scientific knowledge (p. 847). Furthermore, besides the work of knowledgeable scientists, the term "practice" includes the work of ordinary people (e.g., workers and peasants). This provides the background for understanding objectivity as a consequence of everyday practice in different endeavors.

According to Wan, Wong, and Zhan (2013):

> Since Marxists insist on the necessity to understand phenomena from their surrounding conditions, they also believe that science should be understood in its broad social context.

> It is stated that "where would natural science be without industry and commerce?" (Marx & Engels, 1970) .... However, it should be noted that the emphasis on the influence of the social context on scientific activities does not lead Marxism in the anti-rationalism that characterizes various branches in the contemporary philosophy of science. Instead, the social influence on science is just considered as the opposite of and in a unity with rationality or objectivity of science. (p. 1122). Classified as Level III.

It is interesting to note that the two presentations presented above in this section deal with Marxism and still have some subtle differences. Deng et al. (2014) emphasize the importance of practice in Marxism and thus social and cultural influences are sacrificed or ignored as compared to objectivity and rationality in science. On the other hand, Wan et al. (2013) suggest that although the social influence in China is considered less important it is still considered as part of a unity that includes the rationality and objectivity of science.

According to Skordoulis (2008), Epicurus rather than Hegel emerges as the pivotal figure in Marx's early development: "Rather than contained within the idealist philosophy of the Hegelian system, Marx's thesis aimed at formulating an anti-teleological materialism that incorporated the 'activist element' of Hegelianism. Building on Epicurus, Marx's emergent materialism denied neither the objectivity of nature, as Hegel did, nor humans' active relation to nature and to each other" (p. 565). Classified as Level II. Besides pointing out the relevance of objectivity for Marx, this presentation recognizes its importance for Marx due more to the influence of Epicurus rather than Hegel.

3.2.20   Mathematics Education and Objectivity

Patronis and Spanos (2013) have recognized the role of hermeneutics in mathematics education and consider Lakatos's (1976) hermeneutical reconstruction of a historical theme (polyhedral, Euler's formula and related concepts) as an example. Furthermore, they provide the following guideline for classroom practice:

> Setting up a "scene" in the mathematics classroom, with a crucial "opening question" in the beginning, may provide a rich field to initiate a dialogue and give the opportunity for knowledge conflicts and negotiation of meaning. As Skovsmose ... indicates by his examples of project work in the classroom, his reformulation of exemplarity may become a link between educational theory and practice, by planning a thematic approach in mathematics education. We need, however, to explore further the nature of "exemplary themes" in mathematics, which we intend to do now, moving towards a theoretical direction which questions the objectivist trend in mathematics education. (Patronis & Spanos, 2013, p. 1997). Classified as Level III.

As a classroom teaching strategy, Patronis and Spanos (2013) suggest the following sequence: setting up of a scene → opening question → dialogue → conflicts → negotiation of meaning. Indeed, this helps to question the objectivist trend not only in mathematics but also in science education (cf. Lee & Yi, 2013; Niaz, 1995a, b). Daston and Galison (2007) provided similar advice based on the dilemma faced by those who tried to understand electroencephalographs using mechanical objectivity based on "a rigid adherence to rules, procedures, and protocols" (p. 325). Instead, they suggested that the electroencephalographer had to cultivate a new kind of scientific self, one that was more intellectual rather than algorithmic. It is high time that science educators recognize the importance of being "intellectual" in the classroom and ignore algorithmic teaching strategies.

According to Ernest (1991), objectivity of mathematics can be accounted for as socially accepted knowledge, in other words, it is objective by virtue of its acceptance by the scientific community. Rowlands, Graham, and Berry (2011) criticize Paul Ernest's philosophy of mathematics education and defend teaching of mathematics as a formal, academic system of knowledge.

> For Ernest (1991), this is not objectivity in the sense of logical necessity from which the objectivity can be recognised; rather, subjectivity becomes objectivity through consensus. The rationale for this is the failure of the foundationalist programme to establish certainty in the foundations of mathematics: take away the certainty of mathematics then you can take away logical necessity as having any role in establishing what is to be accepted— objectivity merely becomes part of that which is accepted .... What "absolutist" philosophies (Ernest's term for the foundationalist programme) have failed to establish is not logical necessity but absolute certainty in the foundations, but take away logical necessity (because it cannot be "established") and you have objectivity as synonymous with consensus in the sense that they are not separate entities from which the former may play a part in establishing the latter. (Rowlands, Graham, & Berry, 2011, pp. 641–642). Classified as Level III.

Rowlands et al. do recognize the criteria used by Ernest for social acceptance, namely mathematical journals and reviewers. However, in their opinion it is not enough to say that objectivity can be equated with acceptance. Furthermore, in order to support their thesis of how objectivity cannot be equated with acceptance, Rowlands et al. (2011) provide the example of the 4-color theorem. This theorem was proven first by Alfred Kempe in 1879 and later by Peter Tait in 1880. However, 10 years later in 1890 it was found that both "proofs" contained fallacies. This episode led Rowlands et al. (2011) to conclude that consensus for proof (1880–1890) did not mean that the theorem was proved and hence objective. Despite the merit of this interpretation one could argue that it was the community that revealed the fallacies in the theorem and hence shows mathematics to be socially accepted knowledge, as suggested by Ernest (1991). This also illustrates Daston and Galison's (2007) thesis of the evolving nature of objectivity, which is socially conditioned by the scientific community.

Fiss (2012) has analyzed reform movements in mathematics education (based on the documents of the National Education Association, 1894) during the last decades of the nineteenth century that emphasized objective methods of teaching and recommended that rules be derived inductively. Based on this perspective Fiss (2012) concluded:

> This language of objectivity and objects was a novel nineteenth-century reinvention of the scholastic distinctions between subjectivity and the objectivity. At this time, its presence signaled a connection to the physical sciences, as well as a sense of a "scientific self" (Daston & Galison, 2007, pp. 191–252). This language, coupled with the argument that students should use the manipulation of physical objects in the world as a substitute for the epistemic authority of a book or teacher, ultimately reframed mathematics as a physical science (p. 1192). Classified as Level III.

According to Daston and Galison (2007, p. 198), in the mid-nineteenth century the "scientific self" was considered to be an obstacle to mechanical objectivity and following measures were suggested to combat subjectivity: self-restraint, self-discipline, and self-control.

3.2.21   Model of Intelligibility and Objectivity

Drawing on the use of a balance, Machamer and Woody (1994) draw implications for the intelligibility of a model:

> The model exhibits all and only those properties that are important. This intelligibility and the normative character of the idealized model is what allows for objectivity. If a problem cannot be reduced to these elements, or if a participant in the investigation insists on attending to other aspects, then either the problem falls outside the scope of the model or the participant needs (re-)training about what is important in the problem or what are the allowable procedures. Such disagreements can be used to test the scope and adequacy of models, and sometimes give rise to "revolutions" in intelligibility when people become convinced that something important is being left out (p. 224). Classified as Level III.

This illustrates what Machamer and Wolters (2004) later referred to as "both rationality and objectivity come in degrees."

3.2.22   Nature of Science and Objectivity

Nature of science is a controversial topic of considerable interest to science educators and had the following five presentations: Talanquer (2013), Irzik and Nola (2011), Wong, Kwan, Hodson, and Jung (2009), Gauch (2009), and Galili (2011).

Based on the work of philosophers, historians and science educators, Talanquer (2013) has contested the Universalist characterization of the nature of science (NOS) and then concluded:

> The central claim is that scientists in different disciplines have distinctive epistemic goals, practices, and norms that influence how they conduct their research and how they perceive, communicate, and evaluate their activities and results. Their work relies on unique experimental approaches, particular deployments of instrumentation, different forms of explanation, as well as on distinct conceptions of rationality, standards of objectivity, and modes of argumentation. From this perspective, science educators need to better understand what the various practices of the different sciences look like in order to devise more authentic contexts for the teaching and learning of each of these disciplines in schools. (p. 1762). Classified as Level III.

This presentation calls attention for the need to understand diversity in the scientific enterprise. If scientists use unique experimental procedures in order to solve complex problems then their conceptions of rationality, modes of argumentation, and standards of objectivity would also vary accordingly. Precisely, this also characterizes the evolution of objectivity in the history of science.

According to Irzik and Nola (2011), some of the items mentioned in the consensus view of NOS (this generally refers to Lederman, Abd-El-Khalick, Bell, & Schwartz, 2002) lack sufficient systematic unity which leads to a tension among such aspects and then they go on to provide the following example:

> For instance, scientific knowledge is said to be theory-laden and subjective. Does this make objectivity of science impossible? If not, why not? If science is socially and culturally embedded, how is it that it produces knowledge that is valid across cultures and societies? Is the influence of society on science good or bad? How do we distinguish between these two kinds of affects? Does science have any means of detecting the bad ones and eliminating them? These are important questions that need to be raised if we want our students to have a sophisticated understanding of NOS (p. 593). Classified as Level III.

After critiquing the consensus view of NOS (nature of science), Irzik and Nola (2011) then go beyond to assert the objectivity of science as experiments are reproducible and the same experiments done under the same conditions do come up with the same results. This is precisely what Daston and Galison (2007) have referred to as mechanical objectivity. Furthermore, this ignores the fact that in the history of science various scientists doing the same experiments and having the same results came up with entirely different theories. In most parts of the world introductory science courses primarily deal with the history of science and "science in the making." According to Laudan (1996):

> The fact is that scientists do not need to study the history of their discipline to learn the Tradition; it is right there in every science textbook. It is not called history, of course. It is called "science," but it is no less the historical canon for all that. Thus, the budding chemist learns Prout's and Avogadro's hypotheses, and Dalton's work on proportional combinations; he learns how to do Millikan's oil drop experiment; he works through Linus Pauling's struggles with the chemical bond. (p. 153)

It seems that Laudan was writing the science/chemistry curriculum. Furthermore, history of science is replete with controversies among scientists (cf. Machamer, Pera, & Baltas, 2000). This obviously leads to a dilemma: which history shall we include in the classroom? One laden with experimental details or the one based on theory-laden nature of observations leading to controversies in the history of science. History of science bears witness to the difficulties involved in interpreting experimental data and that the essence of the scientific endeavor is perhaps characterized by the creativity and imagination of the scientists. Under this perspective, telling students that scientists are "objective" and "rational" would be too simplistic. It would be more motivating to reconstruct the different historical episodes in order to illustrate "science in the making" and how science is practiced by scientists (Levere, 2006; Niaz, 2012).

Later in the same article, Irzik and Nola (2011) state that scientific knowledge, though theory-laden, is nevertheless reliable because it is obtained by subjecting our theories to critical scrutiny, and

> Similarly, the fact that science is objective (in the sense that scientific findings are correct independently of individual, social and cultural variations) is a result of the same intersubjective critical process. That scientific experiments are reproducible also contribute to the

objectivity of scientific knowledge. Whoever does the same experiment under the same conditions should come up with the same result regardless of when and where the experiment is carried out. Again, it is not clear in the consensus view how reliability and objectivity of science is to be explained without such considerations. (Irzik & Nola, 2011, p. 602)

Nevertheless, this overlooks the fact that some long-standing controversies in the history of science were difficult to resolve and continue to provide considerable difficulties to students' experiences in the lab. An interesting example is the oil drop experiment (Klassen, 2009) which provides, even at present, very contradictory results in almost all parts of the world even with modern apparatus. Daston and Galison (2007) refer to the resolution of the controversy with respect to the oil drop experiment not due to the reproducibility of experimental data, but as an example of "trained judgement." Also with this background consider Martin Perl's philosophy of speculative experiments. Finally, it seems that Irzik and Nola (2011) follow quite closely Kuhn's (1970) advice to science educators, that is just teach "normal science" (for a critical appraisal of Kuhn's "normal science" see, Niaz, 2011, Chap. 2, pp. 17–33).

Wong et al. (2009) turned crisis into opportunity by using the Severe Acute Respiratory Syndrome (SARS) to understand and teach the theory-laden observations as part of nature of science in the classroom. They used an historical account of the "hunt" for the causative agent of SARS that was infused with several examples of theory-laden nature of observations. In one of the video clips they showed that immediately following the announcement on March 18, 2002, by a group of scientists from Hong Kong and Germany that the virus causing SARS was paramyxovirus, other research groups around the world quickly announced that they had also found evidence that paramyxovirus was the causative agent of SARS. Interestingly,

> However, only a few days later, on 22 March 2003, another group of researchers in Hong Kong announced that further evidence showed that coronavirus, rather than paramyxovirus, is the causative agent of SARS. Immediately after this announcement, several laboratories, including Rotterdam, Frankfurt and the Center for Disease Control and Prevention (CDC) in Atlanta, also confirmed the coronavirus theory. This episode illustrates the theory-laden nature of observation and shows how scientists' expectations or predictions influence what they see and how they interpret the data. Acknowledgement of the biased observation of data is in stark contrast to the usual school science curriculum portrayal of scientists as objective and impartial in interpreting data (Wong et al., 2009, p. 110, as part of a section entitled: "Objectivity of scientists and theory-laden observation"). Classified as Level IV.

This episode clearly shows the importance of "science in the making" and how it can facilitate students' understanding of theory-laden nature of observations and that objectivity is an ideal that comes with lot of effort and perhaps only in degrees (cf. Machamer & Wolters, 2004). Furthermore, Wong et al. (2009) consider the initial acceptance of the paramyxovirus as the causative agent of SARS and its replacement by the coronavirus as a consequence of new evidence, as an illustration of the tentativeness of science, which is related to an essential characteristic of good science, such as skepticism and open-mindedness.

According to Gauch (2009):

> The Congress of the United States wanted a current assessment of science's rationality and objectivity, so a 1993 symposium was co-convened by Representative George Brown and the AAAS for the purpose of providing "a philosophical backdrop for carrying out our responsibilities as policymakers" (p. iii). One contributor, influenced by Kuhn, reported that scientists should accept the new picture of science as myth. "Some scientists are still scandalized by the historical insight that science is not a process of discovering an objective mirror of nature, but of elaborating subjective paradigms subject to empirical constraints ... Nevertheless, it is important to understand the nature, function, and necessity of scientific paradigms and other myths ..." (Ronald D. Brunner, in Brown, 1993, p. 6) (pp. 687–688). Classified as Level III.

Gauch (2009) concluded that it is misleading to say that science is tentative, approximate and subject to revision and that some scholars might prefer that policymakers receive a less skeptical and more balanced view of science's powers and limits (p. 688).

Galili (2011) has pointed out the predicament often faced by science educators in understanding and explaining the essence of objectivity. Consider the following statements:

> Thus, the resultant knowledge of *classical mechanics* enabled great technological achievements—a reliable test of objectivity: people walked on the Moon regardless various individual details in the knowledge of the people who created the knowledge required for such enterprise. (p. 1310, original italics)

Many teachers and textbook authors would subscribe to such statements that facilitate an important aspect of the nature of science, namely its objectivity. However, Galili (2011) goes beyond by stating:

> Furthermore, in science education, it is important not to confuse various *aspects* of scientific knowledge with its *genus* .... Confusion of *objectivity* with *universal and unconditional correctness* of knowledge seemingly leads to misconceptions about the nature of science (p. 1310, original italics). Classified as Level IV.

Indeed, conditional correctness of scientific knowledge precisely leads to the evolving nature of objectivity (Daston & Galison, 2007). In other words, just as science advances our understanding also changes, and this shows the need for science educators to understand how objectivity evolves. Indeed, the changing or the tentative nature of scientific knowledge has been recognized as an important part of NOS in many reform documents, and can help to understand objectivity in a historical perspective.

3.2.23 Observation and Objectivity

Sievers (1999) critiques Alan Chalmers' understanding of observation as outlined in his *What is this thing called science?* According to Chalmers when two similar cameras take a picture of the same thing, they produce two identical images.

However, Chalmers argues that when two persons "see" the same thing, there are two different experiences, which may be considered as subjective experiences. Consequently, human beings are unlike cameras as "… an object does not produce in each of us the same subjective experience" (Sievers, 1999, p. 389). After outlining Chalmers position, Sievers (1999) goes on to assert the objectivity of observation in the following terms: "On this view, the objectivity of observation ceases to be a philosophical dogma. We can justify our observations in the face of the subjectivist doubts. In so far as people can be trained to be reliable observers, their perceptual knowledge is objective. Such training is an important part of scientific education" (p. 392). This interpretation in which the objectivity of observation can be restored (based on training) approximates to what Daston and Galison (2007) have referred to as "trained judgment." Those who work in the lab (both students and scientists) can face a dilemma in which they have to make observations, and it is plausible to suggest that "trained judgment" could be one alternative to reach consensus in the case of differences or controversies with respect to the interpretation of data. Classified as Level IV.

Felipe Folque, a prominent figure in the development of astronomy as a discipline in Portugal, taught astronomy and geodesy at the Lisbon Polytechnic from 1837 to 1856. Students received an intensive training in the use of astronomical instruments and mathematical methods that were believed to be important in their future work. Carolino (2012) has summarized this experience in which engineers received training at the Lisbon Polytechnic, in the following terms:

> Historians have stressed the importance that the rise of a culture of precision measurement, from the late eighteenth through the nineteenth century, played in the process of formation of nation-states in Europe and America …. The same happened in nineteenth century Portugal, where the strengthening of a culture of precision and objectivity was especially visible under the reformist government, from mid-nineteenth century onwards. Normalization of methods, standardization of procedures and culture of objectivity guided the work of the technical staff that worked for the General Board for the Geodetic, Chorographic and Hydrographical Works under Folque's direction (pp. 126–127). Classified as Level II, as it refers to objectivity as an academic objective.

This historical experience in the teaching of astronomy and geodesy in the nineteenth century corresponds quite closely to what Daston and Galison (2007) have referred to as "mechanical objectivity."

At this stage it would be interesting to compare the two presentations: Sievers (1999), classified as Level IV, and Carolino (2012), classified as Level II. According to Carolino, students' work was guided by normalization of methods, standardization of procedures and the culture of objectivity. On the contrary, Sievers emphasizes that objectivity is a consequence of training provided to the observers (trained judgment according to Daston & Galison, 2007). Although, both recognize the importance of objectivity, the difference between the two precisely provides an understanding (Sievers) of the evolving nature of objectivity.

In 1860, Herbert Spencer emphasized the importance of science and scientific knowledge. Based on these ideas, Otis W. Caldwell (1869–1947), a botanist and science educator designed general science courses by emphasizing the role played by observations. These courses had considerable popularity in the USA, and according to Heffron (1995), this could be attributed to, "... the historical relationship between science and general education, a relationship established in the opening decades of this century, when the authority of science and scientific objectivity was in the minds of most educators unimpeachable" (p. 227). Next, Heffron (1995) presents a critique of the inductive methods and observations in the following terms:

> If, as Karl Popper and others have argued, science itself does not advance "solely by inductive methods," that is, by the simple stockpiling and ordering of observations, however repetitious, we cannot expect to make our children (often considered "natural scientists" because of their superior observational skills) better scientists by simply making them more observant. We must first make them more theoretical. For in the realm of science, theories come logically before problems, problems before observations. The latter, in so much as they fail to lead to the falsification of these theories, are actually an aspect of non-science. (p. 245). Classified as Level III.

From a Popperian perspective, Heffron has emphasized that the real test of scientific truth lies not in its obedience to our observations, but in its falsifiability, the belief that scientific truths are only temporarily valid and subject ultimately to falsification. Based on this perspective, Heffron concluded that Caldwell's vision of science in general education was fundamentally unscientific and even miseducative (p. 245). Furthermore, it is important to note that Popper's ideas on falsification have been the subject of considerable controversy in the philosophy of science literature (cf. Lakatos, 1970).

3.2.24   Piaget's Epistemic Subject and Objectivity

Piaget's developmental stages have been the subject of considerable controversy in both the psychology and science education literature. Brainerd (1978) has critiqued Piaget's developmental stages on empirical grounds, namely children and adolescents do not acquire the different stages at the ages stipulated by the theory, and hence Piagetian theory has been falsified. This is a very Popperian approach to understand progress and ignores the fact that Piaget's *oeuvre* is based on the *presupposition* that developmental stages correspond to an *epistemic subject*—universal scientific reasoning, ideally present in all human beings (cf. Beth & Piaget, 1966, p. 308). In other words, Piaget was not studying the average of all human abilities, but rather the ideal conditions under which a psychological subject (a particular person) could perhaps attain the competence exemplified by the epistemic subject (for details see Niaz, 1991, p. 570).

Kitchener (1993) has emphasized the important distinction between the epistemic and psychological subject in Piaget's genetic epistemology. In order to understand this distinction he draws on Galilean methodology, a version of the hypothetico-deductive method to indirectly test a hypothesis, in the following terms:

> Since a direct empirical test of his hypothetical law was not possible, he [Galileo] used his famous inclined plane experiment to show that as the angle of incidence approximated 90 °

(free fall), the acceleration of objects rolling down an inclined plane increasingly approximated a constant. Hence, by extrapolation, one may assume it is also true of free fall as a limiting case. Here we have an indirect confirmation of a mathematical law which is true only of ideal objects under ideal conditions, a law to which real objects approximate only to certain degrees. (Kitchener, 1993, p. 142)

Based on this understanding of Galilean methodology, Kitchener provides the following perspective for understanding objectivity:

Knowledge is not to be naively equated with mere belief (or the brute factual existence of a cognitive structure): knowledge has an inescapable *normative* dimension, one concerning concepts like evidence, objectivity, rationality, validity, truth, etc .... These notions are not ... merely *identical* to simple empirical facts like contingencies of reinforcement, nor can they be *replaced* (as in Quine's (1969) naturalistic epistemology) by brute empirical psychological concepts (Kitchener, 1993, p. 141, original italics). Classified as Level II.

Rowell (1993) has endorsed Kitchener's (1993) interpretation of Piaget's epistemic subject and then concluded: "Presumably an epistemic subject would function in this way, but there is considerable doubt that an actual individual would achieve rationality and objectivity in the absence of other social agents (Kitchener, 1981)" (p. 133). Classified as Level III.

It is plausible to suggest that as the epistemic subject does not exist and hence objectivity can only be a possible ideal that can be achieved, provided all the "social agents" required for cognitive development are operative. Kitchener emphasizes that just like validity and truth, objectivity is part of the normative dimension (epistemic subject) and hence cannot be reduced to an empirical psychological dimension (psychological subject). In a sense, both Kitchener (1993) and Rowell (1993) not only recognize the elusive nature of objectivity but also approximate Daston and Galison's (2007) understanding of the evolving nature of objectivity.

3.2.25   *Presuppositions and Objectivity*

School science generally endorses a view that comprises of: (a) Foundationalism, science is built on a foundation of unproblematic true propositions and (b) Logicalism, science has a logical method to determine which of two competing theories is true (McMullin, 1987, p. 50). History of science, however, shows that actual scientific practice is much more complex in which controversies based on the presuppositions of the protagonists play a crucial role. Indeed, controversies play an important role in the dynamics of science, especially before consensus with respect to facts and theories has been achieved (Silverman, 1992, p. 177).

Silverman (1992) has referred to the difficulties involved in understanding science in cogent terms:

Part of the classical perspective of science is that scientists ideally undertake their work without bias or preconception. Objectivity and open-mindedness are indeed integral attributes of science, but not in this naive sense. Rarely does a scientist commence research in the absence of presupposition as to the outcome; objectivity consists not in denying preconceptions, but in the ability to modify beliefs in the light of emerging evidence. Physicist R. A. Millikan, for example, in his autobiography (1950) expresses his initial grave doubts as to the correctness of Einstein's treatment of the photoelectric effect, a remarkable phenomenon in which light seems to collide with electrons as if it were comprised of small hard corpuscles and not waves. To accept a ballistic interpretation of light:

... was clearly impossible, at least for me, particularly in the Ryerson Laboratory where under Professor Michelson's leadership we were working as continuously and familiarly with the wave-lengths of light as with meter sticks .... (Millikan, 1950, p. 66)

In regard to testing Einstein's equation, however, he [Millikan] expected that he would surely prove it false, yet he had to conclude:

I spent ten years of my life testing that 1905 equation of Einstein's, and contrary to all expectations, I was compelled in 1915 to assert its unambiguous experimental verification in spite of its unreasonableness .... (Millikan, 1950, p. 100)

That is objectivity in science. (Silverman, 1992, p. 168, original italics, underline added)

Interestingly, Millikan's opposition to Einstein's hypothesis of lightquanta (despite the acceptance of the photoelectric equation) continued far beyond 1915 and Holton (1999) considers it an irony as it coincides with textbook versions of the experiment. Stuewer (1975, p. 88) goes beyond by considering this adjustment on the part of Millikan as "shocking," considering the fact that even in 1924, in his Nobel Prize acceptance speech, Millikan still questioned Einstein's hypothesis of lightquanta. In a study based on 103 general physics textbooks (published in USA), Niaz et al. (2010a, b) reported that only five mentioned that Millikan's opposition to the quantum hypothesis could be attributed to his prior presupposition and strong belief in the classical wave theory of light. This clearly shows the relationship between how textbooks conceptualize objectivity and the practice of science based on logicalism (McMullin, 1987).

With respect to the determination of the elementary electrical charge there was a bitter controversy between two protagonists (R. A. Millikan and F. Ehrenhaft), and Silverman (1992) recounts this historical episode by considering that: (a) Study of this controversy helps illuminate subtle and complex issues underlying the experimental interrogation of nature; (b) One does not, as often implied by an idealized perspective of science, simply turn on the apparatus, make measurements, and compare with theory; and (c) Questions always arise over such mundane, yet critical, matters such as the sensitivity of apparatus, effects of systematic and random noise, environmental influences, and the reliability and admission of data. Based on these considerations, Silverman (1992) suggested: "How these questions are answered depends on the philosophical attitudes of the experimenter. Millikan scrutinized his measurements to determine where a particular experimental run was 'good'—that is in keeping with his expectations [elementary electrical charge, electron]. Ehrenhaft accepted all measurements in the belief [fractional charges, sub-electrons] that that constituted objective observation. The general philosophical climate of the experimenters' milieu also played an important role" (p. 169). Classified as Level IV.

Again, general chemistry and physics textbooks (published in USA) completely ignore the presuppositions of both Millikan and Ehrenhaft (for details see Niaz,

2009, Chap. 7). No wonder, neglecting the role played by presuppositions leads textbooks to endorse what Daston and Galison (2007) have referred to as "mechanical objectivity." Silverman's (1992) conceptualization that, objectivity consists not in denying preconceptions, but in the ability to modify beliefs in the light of emerging evidence—provides not only insight into the dynamics of scientific progress but also approximates to what Daston and Galison (2007) have referred to as "trained judgement."

3.2.26   Quantum Mechanics and Objectivity

According to Hadzidaki (2008a), the understanding of objectivity varies in classical physics from quantum mechanics. For example, in quantum mechanics it is not possible to "… interpret the statements of physics as informing us directly of attributes of the entities under investigation—or, in other words, to judge the objectivity of our knowledge through a comparison with the reality per se …" (p. 69). Consequently, only a "weak" form of objectivity based on inter-subjective agreement can be invoked. Classified as Level III.

In a section entitled "objectivity and subjectivity," Pospiech (2003) noted: "Perhaps one of the deepest consequences of uprising quantum theory was the insight that physical truth is not absolute as many people believed after the overwhelming success of Newton's work. Suddenly there seemingly occurred quantum jumps; results could by principle only be predicted with probability and depended on the acting of an observer. Attempts to explain these phenomena in classical terms were frustrating. The concept of fixed properties independent of any measurement for single quantum objects had to be abandoned. Only the result of many equal measurements on equal objects could be predicted and reproduced" (p. 568). Classified as Level III.

3.2.27   Romantic Science and Objectivity

Romanticism as a movement emerged in Germany and spread to Europe in the late eighteenth and early nineteenth century and has been viewed as a cultural and intellectual movement that countered rationalism then considered as the dominant *Weltanschauung* (cf. Cunningham & Jardine, 1990). According to Hadzigeorgiou and Schulz (2014):

> The Romantics gave great importance not only to social and political education—since it was through education that human beings became human and a citizen—but also to science, neither of which is well known or typically associated with romanticism. It was "Romantic science," in fact, while being the development that grew in reaction to eighteenth century Enlightenment rationalism, with its allied mechanistic philosophy (based on objectivity and determinism) that succeeded in actually transforming the latter by emphasizing imaginative/creative thinking and public excitement about scientific work and discoveries … (pp. 1965–1966). Classified as Level III.

According to the authors, given the pragmatist/utilitarian conception of school science prevalent today, romantic science can in contrast provide food for thought by emphasizing the notion of wonder and the poetic/non-analytical mode of knowledge.

3.2.28 Science in the Making and Objectivity

Nielsen (2013) draws attention to the importance of science as a mode of communication that sustains knowledge. Communication among scientists is what makes knowledge possible, namely technical language, rhetorical resources, peer reviews among others. Consequently, without communication perhaps there would be no science:

> Decisions about the topic and resources of ongoing scientific communication involves distinguishing between what Bruno Latour (1987, p. 4) calls "ready-made science," that is, stabilized scientific knowledge in textbooks, and "science-in-the-making," that is, scientific knowledge discussed and negotiated in labs, peer reviews, etc. The implication that there is a close connection between the content and the media of (more or less tentative) scientific knowledge is important to our purposes: It is essential for science learners to realize that, despite the appeals to (absolute) objectivity and universality, scientific knowledge does not exist in and of itself; its tentativeness, or its degree of existence, to put it the Latourian way, depends on the ways in which it is involved in scientific communication (Nielsen, 2013, p. 2082). Classified as Level III.

With this background Nielsen (2013) suggests that the following be included as an eighth item of Lederman's (2007, pp. 833–835, also known as the Lederman seven) list of nature of science topics: science is a mode of communication that enables and sustains knowledge in certain ways (p. 2081). This leads us to understand better the distinction between "ready-made science" and "science-in-the-making." Ready-made science, of course, refers to stabilized scientific knowledge as presented generally in textbooks. It is plausible to suggest that the communicative structure of science would improve if we discuss in class some of the controversial aspects of "science-in-the-making" and how scientists resolved the controversies. Interestingly, this facet of "ready-made science" is widespread in most parts of the world (cf. Niaz, 2016, Chap. 4, in the context of presentation of atomic models in textbooks).

3.2.29 Science, Religion and Objectivity

Based on his criticism of Good (2001) and Mahner and Bunge (1996), with respect to the religious habits of mind, Gauld (2005) has called for a careful scrutiny of the writings of Christian scientists (e.g., Polkinghorne):

> In the above discussion it has been argued that, when one considers a wider range of evidence than Good (2001) has done, the scientific and religious habits of mind are more

similar to one another than he acknowledges. In both cases openness to argument and evidence, skepticism, rationality and objectivity are all held in high regard; in both some ideas are more protected from attack while others are more open to challenge; and in both, at any time, there are various degrees of commitment to theories from skeptical rejection to passionate endorsement. Both habits of mind stem from the same scholarly attitude and any difference between them is probably due to differences in what are counted as appropriate evidence and good reasons. For example, in the Christian religion historical evidence and evidence from human agency and self-awareness are more important than they apparently are in physics (pp. 301—302). Classified as Level II.

This is an interesting example of considering objectivity in scientific and religious habits of mind as academic objectives. Furthermore, it can facilitate a better understanding of both religion and science and also help in teaching controversial topics of the science curriculum, such as evolution.

According to Pennock (2010):

IDC [Intelligent Design Creationism] shows in a striking manner how radical postmodernism undermines itself and its own goals of liberation. If there is no difference between narratives—including no difference between true and false stories and between fact and fiction—then what does liberation come to? Are scientific investigations of human sexuality really no more likely than the Genesis tale of Eve's creation from Adam's rib? Those original goals—the overthrow of entrenched ideologies that hid and justified oppression—that motivated the postmodern critique were laudable. But the right way to combat oppression is not with a philosophy that rejects objectivity and relativizes truth, for that guts oppression of its reality (p. 777). Classified as Level III.

Pennock is arguing that the post-modern rejection of objectivity is double edged: on the one hand it espouses liberation from different forms of power structures and at the same time it provides IDC an argument against the prestige of objectively determined knowledge provided by science. Proponents of IDC have acknowledged that it is precisely for this reason that they consider themselves to be deconstructionists and postmodern (cf. Pennock, 2010, p. 759). In this context, it would be helpful to consider some of the ideas introduced by Gauld (2005) with respect to openness to argument and evidence in both science and religion.

3.2.30 Scientific Literacy and Objectivity

According to Krogh and Nielsen (2013), in order to achieve functional literacy, "… it is necessary to help students dismantle the naïve view that science is objective and value free, and give the more realistic impression that objectivity is not an all or nothing thing. There are degrees of objectivity" (p. 2061). Classified as Level III. Furthermore, the authors suggested that the inclusion of recent debates within the scientific community based on discipline-specific NOS-insights can help students to understand this facet of science. Machamer and Wolters (2004) have presented a similar thesis with respect to degrees of objectivity.

3.2.31 Scientific Method and Objectivity

Based on a framework that emphasizes the technological dimension, Gil-Pérez et al. (2005) have referred to the wide-spread practice in science education of associating objectivity with the scientific method:

> For example, in interviews held with teachers, a majority have referred to the "Scientific Method" as a sequence of well defined steps in which *observations* and *rigorous experiments* play a central role which contributes to the *exactness* and *objectivity* of the results obtained. Such a view is particularly evident in the evaluation of science education: as Hodson (1992) points out, the obsessive preoccupation with avoiding ambiguity and assuring the reliability of the evaluation process distorts the nature of the scientific approach itself, essentially vague, uncertain, intuitive (p. 313, italics in the original). Classified as Level III.

Indeed, the ambiguity, uncertainty, creativity, and intuitive aspects of the scientific endeavor are essential if we want our students to understand "science in the making."

Depew (2010) has referred to the scientific method in the context of Darwinism:

> Ironically, so well has the folk version of simplistic empiricism about "scientific method" been internalized into the post-Sputnik public sphere that, rather than reading Kuhn's Structure of Scientific Revolutions as attacking this view of scientific method, students usually read it as expressing mere skepticism about the scientific objectivity with whose norms they are already familiar. Nor do many post-Kuhnian social constructionists do anything [to] counter this impression. In fact, some of them actually play into it. Under such conditions, portraying evolutionary science in any way that seems not to fit the model of well-confirmed science whose rudiments people, including journalists, learned in school generates in most audiences not a more complex conception of scientific inquiry suited to an inherently complex subject, but a sense that Darwinism is not really a science at all, but instead a world view or secular religion (pp. 361–362). Classified as Level III. This presentation could have been classified in Evolution, creationism and objectivity.

It is important to note the difficulties involved in teaching evolution and how at times Darwinism is not considered really a science but perhaps a secular religion. Indeed, to promote the idea that all science is well confirmed is misleading and the inability to discuss this in class leads to the difficulties involved in teaching evolution and understanding Darwinism.

According to Kosso (2009): "The point here is that the scientific method, and the information gained through observation, can be essentially under the influence of what the scientists have in mind, without compromising the objectivity of the method or the information" (p. 38). Classified as Level I. Kosso's argument is that scientific method is essentially global, in other words any model that describes testing of individual hypotheses, one at a time and in isolation from other theoretical information, is inaccurate (p. 41). However, textbooks generally argue that it is a sequence of steps in a scientific method that makes science objective and this creates difficulties in understanding how science is done.

3.2.32   Scientific Methodology and Objectivity

Rusanen and Pöyhönen (2013) have suggested that scientific concepts could be understood as communally shared epistemic tools that scientists use to coordinate their efforts in their common tasks of knowledge production. Working with mechanisms of conceptual change, these authors have reported that: "… the objectivity and correctness of scientific inference are guaranteed by communication and error correction within the research group and within the wider scientific community. Importantly, this picture of scientific concepts applies also in less strongly distributed cases: what is referred to by speaking of scientific concepts are not mental representations of individuals but pieces of scientific knowledge that can be shared by a community of individuals" (p. 1393). Classified as Level IV. This presentation approximates to Daston and Galison's (2007) idea of "trained judgment."

Develaki (2008) first points out that the traditional ethics of science are based on objectivity, empirical control, and precision measurement. Furthermore, scientific knowledge is also projected as autonomous and neutral since it was considered to be substantiated and established exclusively on the basis of empirical and logical criteria. In contrast, critical philosophy focuses on the interaction between science and society:

> The view that the evaluation and choice of theories is based (solely) on unambiguous logical rules and empirical criteria has been challenged on the grounds that the development and choice of theories takes place under the deciding influence of concrete world views (e.g., a mechanistic world view for classical mechanics), so that the resulting incommensurability of theoretical and methodological standards of the various theories precludes a neutral, objective and fair framework for comparison and selection among alternative theories (e.g., Toulmin, Hanson, Bohm, Kuhn, and others, see in Suppe, 1977). (Develaki, 2008, p. 875). Classified as Level III.

Comparing the presentations of Rusanen et al. (2013) and Develaki (2008), it can be observed that the former explicitly posits the critical role played by communication within the scientific community, whereas the latter only refers to the problematic nature of objectivity.

3.2.33   Scrutinized Scientific Knowledge and Objectivity

Abd-El-Khalick (2013) has clarified the difference between the social and relativistic notions of scientific knowledge in the context of understanding objectivity:

> The social NOS, or "science as social knowledge," refers to the epistemic function of these social activities: It refers to the constitutive values associated with those established venues for communication and criticism within the scientific enterprise (e.g., blind review processes), which serve to enhance the objectivity of collectively scrutinized scientific knowledge through decreasing the impact of individual scientists' idiosyncrasies and subjectivities (Longino, 1990). In this specific sense, it should be noted, social NOS refers to

conceptions of science as advanced by philosophers of science such as Helen Longino ... and should *not* be confused with relativistic notions of scientific knowledge (p. 2096). Classified as Level IV.

Ford (2008) has referred to the dilemma faced by a scientist during theory choice, as no set of objective rules can provide guidelines for selecting a theory:

> However, it is becoming clear not only in the science studies literature but also in psychology that the information provided by any set of rules or method (i.e., declarative knowledge) is insufficient to account for inquiry. For example, Machamer and Osbeck (2003) elaborated on this point in light of Kuhn's account of how scientists choose among rival theories, noting that no set of objective rules can explain theory choice sufficiently. The key insight offered by Machamer and Osbeck (2003) is that one also needs to know under what circumstances and in what way (and, indeed, it seems, to what end) any posited rules should be applied so their application is appropriate (p. 152). Classified as Level III.

3.2.34   Social/Cultural Milieu and Objectivity

According to Cobern (1995):

> Colloquial positivism roughly represents a classical view of realism, philosophical materialism, strict objectivity, and hypothetico-deductive method. Though recognizing the tentative nature of all scientific knowledge, colloquial positivism imbues scientific knowledge with a Laplacian certainty denied all other disciplines, thus giving science an a priori status in the intellectual world (p. 299). Classified as Level III.

By colloquial positivism Cobern (1995) is not referring to the philosophical sense, generally referred to as logical positivism or logical empiricism, but rather in the sense of a mythology of school science as referred to by Smolicz and Nunan (1975). Based on this clarification, Cobern (1995) then goes on to critique the traditional practice of science education:

> While it may never have been explicit, the goals of science education clearly have been to persuade students that science provides a fairly constant, highly justified, and sufficient understanding of physical phenomena .... The claim of certainty for scientific knowledge which science educators grounded in positivist philosophy was rendered untenable years ago and it turns out that social and cultural factors surrounding discovery may be at least as important as the justification of knowledge. (p. 287)

Cobern' main concern is to show that discovery in science inevitably takes place in a social and cultural milieu and lacks the certainty school science tries to convey as a dogma (cf. as reproduced in Niaz, Klassen, McMillan, & Metz, 2010b). Interestingly, a recent study has highlighted the importance of the status of certainty/uncertainty of physics knowledge as a means to facilitate conceptual understanding: "The knowledge that has already been acquired allows the researchers to raise new questions because there is uncertainty; a given study aims to decrease this uncertainty and then new questions emerge, again pointing out new uncertainty. This dynamics of uncertainty based on knowledge is a way of developing knowledge. We also consider that, in the students' processes of construction of knowledge, uncertainty can drive the learning process of knowledge" (Tiberghein, Cross, & Sensevy, 2014, p. 931). This clearly shows that uncertainty with respect to scientific knowledge need not be a constraint in learning science but rather can even facilitate construction of new knowledge. Consequently, questioning the role of objectivity in the "strict" sense has important implications for science education.

3.2.35   *Social Nature of Scientific Knowledge*

According to Howard (2009), "science's own unreflected pretensions to objectivity" (p. 212) needs to be countered with the social dimensions of knowledge as reflected in the early work (Mannheim, Fleck, Zilsel, & Merton) and more recent work on the social epistemology of science (Longino, Solomon, & Kusch). However, he feels that work on the social dimensions of scientific knowledge has been somewhat peripheral to mainstream work in epistemology and philosophy of science, and that the field has yet to mature. For example, Howard considers (p. 212) Steve Fuller's intervention unfortunate on behalf of the defendants, hence on behalf of requiring the teaching of intelligent design in public schools, in the Katzmiller v. Dover case of 2005. Fuller was the founding editor of the journal *Social Espistemology*, that aspired to be an effective voice in the reform of scientific and social practice affecting science. Classified as Level III.

According to Uebel (2004): "Yet note that the [Vienna] Circle's intersubjective meaning criterion did not only play a negative but also a positive role (it was not merely an ad hominem device for segregating metaphysics). The notion of intersubjectivity also provided the framework within which it was possible for science to attain its autonomy from philosophy: it opened the possibility for replacing the 'metaphysical' idea of objectivity. The objectivity of science did not consist in the provision of distortionless reflections of reality—of 'views from nowhere'—but in the possibility for intersubjective control of perspectival views and assertions" (p. 54). (Classified as Level III). Based on these considerations, Uebel concluded that the intersubjective perspective required not only the adoption of radical fallibilism but also the recognition of the social character of scientific knowledge.

According to Allchin (1999): "The many cases of bias and error in science have led philosophers to more explicit notions of the social component of objectivity. Helen Longino (1990), for example, underscores the need for criticism from alternative perspectives and, equally, for responsibly addressing criticism. She thus postulates a specific institutional, or social, structure for achieving Merton's 'organized skepticism'" (p. 6). Classified as Level III.

It can be observed that the science education literature has shown considerable interest in the social nature of scientific knowledge and consequently its implications for classroom practice, especially for teaching controversial topics.

3.2.36   Theory-Laden Observation and Objectivity

Based primarily on the work of Kuhn (1970) and the Duhem-Quine thesis, observations are influenced by the theories/beliefs one holds. In other words all observations are based on some essential theoretical assumptions that may influence the degree to which a scientist may be objective (Godfrey-Smith, 2003). Based on this background, Lau and Chan (2013) designed a study (based on the conceptual change model of Hewson, Beeth, & Thorley, 1998; Posner, Strike, Hewson, & Gertzog, 1982) to explore the effect of theory-laden observations on students understanding of a lab activity:

> A *discrepant event*, the manipulated theory-laden observation, is used to create *cognitive conflicts* on students' beliefs about the objectivity of observation and science. Then students' practical epistemologies are worked on publicly and explicitly through dialogue, by which the conceptions of theory-ladenness is made *intelligible* and *plausible* to students, and as such, conceptual change regarding their formal epistemologies would be likely (Lau & Chan, 2013, p. 2644, original italics). Classified as Level IV.

The lab activity asked students (Grade 9 students in Hong Kong) to investigate whether heating can destroy the vitamin C contents of vegetables. One group of students was told that scientists had found that vitamin C cannot be destroyed by heating and another group was told that vitamin C would be destroyed at high temperature. Lau and Chan (2013) provided the rationale of their study as:

> In such way, the students were "biased" by the two theories in opposite directions in the observation of the end points and/or the report of data. But actually the two vitamin C solutions provided are both unboiled! To make certain if the students had really been convinced by the "theory" given in the task sheet, they were asked to *predict* the results before conducting the experiment. About 83% of them made predictions in line with the "theory" given. (p. 2646, original italics)

Results obtained showed that the two groups of students obtained data in line with the predictions from the given "theories" about vitamin C, which shows the role played by theory-laden observations. These results helped the students to understand the idea that observations cannot be entirely objective. Interestingly, some students thought that they were "tricked" by the instructor and one student expressed, "How come you give us something wrong ..." (p. 2650). Finally, most students became more receptive to the idea that observations are not truly objective. Designing such studies can be helpful in facilitating a better understanding of the scientific endeavor.

The role of theory-laden observations and objectivity has been the subject of a study by Park, Nielsen, and Woodruff (2014). On the one hand, these authors recognize the importance of theory-laden observations but still recognize its problematic nature: "Popper ... partially endorsed the notion of theory-free observation when a radical change of theory occurs because past experiences or theories cannot guide scientists to modify the anomalies; rather, objectivity, rationality and elimination of subjectivity lead to new theory. Einstein ..., Heisenberg ..., and Feynman ..., outstanding physicists argue that neither 100% theory-independent, nor 100% theory-dependent observation really exists" (p. 1172). Later, in this context these authors illustrate their thesis by providing the example of observations provided by the 1919 eclipse experiments: "Without observational and empirical evidence, a theory cannot stand. For instance, when Einstein suggested the special theory of relativity in 1915, he was not a famous physicist at all. After the obser-vation of the 1919 solar eclipse by Eddington, Einstein's theory was accepted and then, Einstein became famous" (p. 1172). The actual events related to the eclipse experiments were much more complex. Niaz (2009, Chap. 9, pp. 127–137) has argued that if Edington (considered to be a major expert on Einstein's theory of relativity) had not been aware of the theory, it would have been extremely difficult to interpret observations from the eclipse observations, as providing support for the theory. Classified as Level III.

According to Develaki (2012): "In the philosophy, history and sociology of science was developed a series of documented arguments and disputes that challenged the objectivity of observations and the interpretations of experimental data for principal reasons (and also for practical reasons such as the technological insufficiency of the experimental arrangements), which was noted very early (1928 by Duhem): concretely, given their theory-ladenness and theory-guidedness, experiments cannot, or at least cannot always, identify the erroneous hypothesis within the complex interweaving of auxiliary hypotheses and theoretical principles that lead to a specific prediction that is under examination (e.g. Hanson …; Suppe …; Duhem …; Hume …; Popper …)" (p. 867). Classified as Level III. Later Develaki compares the positions of Kuhn, Lakatos, and Giere with respect to theory choice (p. 870) and concludes that only in very favorable circumstances theories are based entirely on logical and experimental grounds.

3.2.37 Values and Objectivity

According to Cordero (1992), scientific practice presupposes both theories and values, which does not necessarily destroy objectivity (p. 50). He then goes on to illustrate scientific practice by exploring the intricate relationship between facts and values: "If history shows anything, it is that in science the facts have rarely been loyal to the values which initially led to their identification. When Darwin developed his theory of evolution, he made liberal use of facts that had been gathered by his teleologically oriented predecessors, but he did not respect the valuations which those facts originally carried. In fact, Darwin's approach turned teleological biology on its head and initiated the destruction of the man-centered and goal-oriented biology then prevalent" (pp. 53–54). According to Cordero this shows the invariance of scientific facts to value change. This, however, may constitute a dilemma for a science educator who believes that science and the values on which it is based are generally objective. Cordero (1992) resolves the dilemma in the following terms:

> The way in which science has forged the objectivity of its values is, I suggest, of particular interest to a certain type of person in the contemporary world. I have in mind a person who

agrees that science is acceptably objective, and who cannot honestly take as legitimate any absolute truths or values, let alone ones that are imposed by mere authority. I am referring to a person that has outlived the quest for absolutes, yet one who is aware of his needs and who has managed to develop a sense of reliable access to the world through scientific thought, however limited this kind of access might look relative to previous "philosophical" or "religious" standards. I will call this person the "humane naturalist" (p. 65). Classified as Level III.

Thus a "humane naturalist" would accept science to be objective and at the same time question absolute truths or values—which reflects the problematic nature of objectivity.

Several feminist philosophers, including Elizabeth Anderson, Helen Longino, and Janet Kourany, have argued that feminist values can help increase the objectivity and rationality of scientific reasoning, including decisions about which theories to accept or reject. Based on this premise, Intemann (2008) has concluded:

If feminist (or any social, ethical, or political values) can play a legitimate role in scientific reasoning, then we must not continue to represent science as "value-free" in science education. We must develop more nuanced and sophisticated accounts of concepts such as "bias," "objectivity," and "scientific rationality" that reflect the complex interactions between science and values (p. 1078). Classified as Level III.

According to Davson-Galle (2012): "…I will contend that, although science is not and cannot be totally value free, the inescapably involved values are benign, not in the sense that that involvement is not influential but in the sense that it does not affect science's status as objective" (p. 192). Lack of a critical perspective may lead many science educators to agree with this interpretation of values in science. Classified as Level II.

After considering the events related to the Vietnam War and the Civil Rights Movement in the USA in the 1960s, Cobern and Loving (2008) have referred to the difficulties involved in understanding objectivity in science, especially in the educational context:

Television brought the war home as people saw for the first time the effects of Napalm, Agent Orange and other products of scientific knowledge in the service of political and military needs. Students in particular were prone to change their estimation of science because of what they perceived as an unholy alliance between the community of science and a military-industrial complex that developed and produced such weapons. The rhetoric of values neutrality and objectivity was not tenable when the science community having taken credit for such things as the Green Revolution now denied any responsibility for Agent Orange and Napalm. Science not only lost its luster, it lost its innocence (p. 431). Classified as Level III.

This presentation highlights the underlying tension between scientific progress and the assumptions with respect to its neutrality and objectivity. It is not difficult to see how for a critical student dissonance may lead to tragedy. In order to grapple with such thorny issues science educators will have to reconsider the traditional values associated with the objective nature of science.

This chapter provides examples of research reported in the journal *Science & Education* (35 sections) that facilitate a wide range of perspectives with respect to understanding objectivity. These examples provide a glimpse of research conducted in various parts of the world over a period of more than 20 years. Conclusions based on these findings along with those of Chaps. 4–6 will be presented in Chap. 7.

References

Atkinson, P., & Delamont, S. (2005). Analytic perspectives. In N. K. Denzin & Y. S. Lincoln (Eds.), *The Sage handbook of qualitative research*. 3rd ed (pp. 821–840). Thousand Oaks: Sage.

Bell, J. S. (1987). *Speakable and unspeakable in quantum mechanics*. Cambridge: Cambridge University Press.

Beth, E. W., & Piaget, J. (1966). *Mathematical epistemology and psychology*. Dordrecht: Reidel.

Blake, D. D. (1994). Revolution, revision or reversal: genetics-ethics curriculum. *Science & Education, 3*(4), 373–391.

Brainerd, C. J. (1978). The stage question in cognitive-developmental theory. *Behavioral and Brain Sciences, 2*, 173–213.

Brown, G.E. (1993). *The objectivity crisis: rethinking the role of science in society*, Chariman's report to the Committee on Science, Space, and Technology, House of Representatives, 103rd Congress, first session, serial D. Washington: U.S. Government Printing Office.

Campbell, D. T. (1988a). Can we be scientific in applied social science? In E. S. Overman (Ed.), *Methodology and epistemology for social science* (pp. 315–333). Chicago: University of Chicago Press. (first published in 1984).

Charmaz, K. (2005). Grounded theory in the 21st century: applications for advancing social justice studies. In N.K. Denzin & Y.S. Lincoln (Eds.), *The Sage handbook of qualitative research* 3rd ed., pp. 507–535. Thousand Oaks, CA: Sage Publications.

Collins, H. M., & Evans, R. (2002). The third wave of science studies: Studies in expertise and experience. *Social Studies of Science, 32*, 235–296.

Cunningham, A. & Jardine, N. (Eds.) (1990). *Romanticism and the sciences*. Cambridge: Cambridge University Press.

Cushing, J. T. (1989). The justification and selection of scientific theories. *Synthese, 78*, 1–24.

Cushing, J. T. (1995). Hermeneutics, underdetermination and quantum mechanics. *Science & Education, 4*(2), 137–147.

Daston, L., & Galison, P. L. (1992). The image of objectivity. *Representations, 40*(special issue: seeing science), 81–128.

Daston, L., & Galison, P. (2007). *Objectivity*. New York: Zone Books.

Deng, F., Chai, C. S., Tsai, C.-C., & Lin, T.-J. (2014). Assessing South China (Guangzhou) high school students' views on nature of science: a validation study. *Science & Education, 23*, 843–863.

Denzin, N. K., & Lincoln, Y. S. (2005). Introduction: the discipline and practice of qualitative research. In N. K. Denzin & Y. S. Lincoln (Eds.), *The Sage handbook of qualitative research*. 3rd ed (pp. 1–32). Thousand Oaks: Sage.

Dewey, J. (1925/1983). Science, belief and the public. In J. A. Boydston (Ed.), *John Dewey: the middle works, 1899–1924 (Vol. 15)*. Carbondale: Southern Illinois University Press.

Ernest, P. (1991). *The philosophy of mathematics education*. London: Falmer.

Feyerabend, P. (1975). *Against method: outline of an anarchist theory of knowledge*. London: New Left Books.

Gergen, K. J. (1994). The mechanical self and the rhetoric of objectivity. In A. Megill (Ed.), *Rethinking objectivity*. Durham: Duke University Press.

Giere, R. N. (2006a). Perspectival pluralism. In S. H. Kellert, H. E. Longino & C. K. Waters (Eds.), *Scientific pluralism* (pp. 26–41). Minneapolis: University of Minnesota Press.

Giere, R. N. (2006b). *Scientific perspectivism*. Chicago: University of Chicago Press.

Glaser, B. G., & Strauss, A. L. (1967). *The discovery of grounded theory*. Chicago: Aldine.

Godfrey-Smith, P. (2003). *Theory and reality: an introduction to philosophy of science*. Chicago: University of Chicago Press.

Good, R. (2001). Habits of mind associated with science and religion: Implications for science education. In W.F. McComas (Ed.), Proceedings of the 6th international history, philosophy and science teaching group meeting. Denver.

Harding, S. (1987). The science question in feminism. Ithaca: Cornell University Press.

Hewson, P. W., Beeth, M. E., & Thorley, N. R. (1998). Conceptual change teaching. In B. J. Fraser & K. G. Tobin (Eds.), *International handbook of science education*. Dordrecht: Kluwer.

Hodson, D. (1992). Assessment of practical work: some considerations in philosophy of science. *Science & Education, 1*(2), 115–144.

Holton, G. (1978a). Subelectrons, presuppositions, and the Millikan-Ehrenhaft dispute. *Historical Studies in the Physical Sciences, 9*, 161–224.

Holton, G. (1978b). *The scientific imagination: case studies*. Cambridge: Cambridge University Press.

Holton, G. (1999). R.A. Millikan's struggle with the meaning of Planck's constant. *Physics in Perspective, 1*, 231–237.

Irzik, G. (2013). Introduction: commercialization of academic science and a new agenda for science education. *Science & Education, 22*(10), 2375–2384.

Keller, E. F. (1985). Reflections on gender and science. New Haven: Yale University Press.

Keller, E. F. (1992). *Secrets of life - secrets of death: essays on language, gender and science*. New York: Routledge.

Kitchener, R. F. (1981). Piaget's social psychology. *Journal for the Theory of Social Behavior, 11*, 253–278.

Kitcher, P. (1993). *The advancement of science: science without legend, objectivity without illusions*. New York: Oxford University Press.

Klassen, S. (2006). A theoretical framework for contextual science teaching. *Interchange, 37*, 31–62.

Klassen, S. (2009). Identifying and addressing student difficulties with the Millikan oil drop experiment. *Science & Education, 18*, 593–607.

Kuhn, T. (1962). *The structure of scientific revolutions*. Chicago: University of Chicago Press.

Kuhn, T. (1970). *The structure of scientific revolutions*. 2nd ed Chicago: University of Chicago Press.

Lakatos, I. (1970). Falsification and the methodology of scientific research programs. In I. Lakatos & A. Musgrave (eds.), *Criticism and the growth of knowledge* (pp. 91–195). Cambridge: Cambridge University Press.

Lakatos, I. (1971). History of science and its rational reconstructions. In R. C. Buck & R. S. Cohen (Eds.), *Boston studies in the philosophy of science Vol. 8* (pp. 91–136). Dordrecht: Reidel.

Lakatos, I. (1976). *Proofs and refutations: the logic of mathematical discovery*. Cambridge: Cambridge University Press.

Latour, B. (1987). *Science in action*. Cambridge: Harvard University Press.

Laudan, L. (1984). *Science and values: the aims of science and their role in scientific debate*. Berkeley: University of California Press.

Laudan, L. (1996). *Beyond positivism and relativism: theory, method and evidence*. Boulder: Westview Press (Division of HarperCollins).

Lederman, N. G. (2007). Nature of science: past, present, and future. In S. K. Abell & N. G. Lederman (Eds.), *Handbook of research on science education* (pp. 831–879). Mahwah: Lawrence Erlbaum.

Lederman, N. G., Abd-El-Khalick, F., Bell, R. L., & Schwartz, R. (2002). Views of nature of science questionnaire: toward valid and meaningful assessment of learners' conceptions of nature of science. *Journal of Research in Science Teaching, 39*, 497–521.

Lee, G., & Yi, J. (2013). Where does cognitive conflict arise from? The structure of creating cognitive conflict? *International Journal of Science and Mathematics Education, 11*, 601–623.

Lefsrud, I. M., & Meyer, R. E. (2012). Science or science fiction? Professionals' discursive construction of climate change. *Organization Studies, 33,* 1477–1506.

Levere, T. H. (2006). What history can teach us about science: theory and experiment, data and evidence. *Interchange, 37,* 115–128.

Longino, H. E. (1990). *Science as social knowledge: values and objectivity in scientific inquiry.* Princeton: Princeton University Press.

Longino, H. E. (2002). *The fate of knowledge.* Princeton: Princeton University Press.

Longino, H. E. (2008). Philosophical issues and next steps for research. In R. A. Duschl & R. E. Grandy (Eds.), *Teaching scientific inquiry: recommendations for research and implementation* (pp. 134–137). Rotterdam: Sense Publishers.

Machamer, P., Pera, M., & Baltas, A. (2000). Scientific controversies: an introduction. In P. Machamer, M. Pera & A. Baltas (Eds.), *Scientific controversies: Philosophical and historical perspectives* (pp. 3–17). New York: Oxford University Press.

Machamer, P., & Wolters, G. (2004). Introduction: science, values and objectivity. In P. Machamer & G. Wolters (Eds.), *Science, values and objectivity* (pp. 1–13). Pittsburgh: University of Pittsburgh Press.

Mahner, M., & Bunge, M. (1996). Is religious education compatible with science education? *Science & Education, 5*(2), 101–123.

Mao, Z. D. (1986). *Selected works of Mao Zhe Dong.* Being: People's Publishing House.

Marx, K., & Engels, F. (1970). *The German ideology.* London: Lawrence & Wishert.

McMullin, E. (1987). Scientific controversy and its termination. In H. T. Engelhardt, Jr. & A. L. Caplan (Eds.), *Scientific controversies* (pp. 49–51). Cambridge: Cambridge University Press.

Merton, R. K. (1979). The normative structure of science. In R. K. Merton (Ed.), *The sociology of science: theoretical and empirical investigations* (pp. 267–278). Chicago: University of Chicago Press.

Midgley, M. (1985). *Evolution as religion.* London: Methuen.

Millikan, R. A. (1950). *The autobiography of Robert A. Millikan.* New York: Prentice Hall.

National Education Association. (1894). *Report of the committee of ten on secondary and school studies, with the reports of the conferences arranged by the committee.* New York: American Book Company.

Niaz, M. (1991). Role of the epistemic subject in Piaget's genetic epistemology and its importance for science education. *Journal of Research in Science Teaching, 28,* 569–580.

Niaz, M. (1995a). Cognitive conflict as a teaching strategy in solving chemistry problems: a dialectic-constructivist perspective. *Journal of Research in Science Teaching, 32,* 959–970.

Niaz, M. (1995b). Progressive transitions from algorithmic to conceptual understanding in student ability to solve chemistry problems: a Lakatosian interpretation. *Science Education, 79,* 19–36.

Niaz, M. (2009). *Critical appraisal of physical science as a human enterprise: dynamics of scientific progress.* Dordrecht: Springer.

Niaz, M. (2011). *Innovating science teacher education: a history and philosophy of science perspective.* New York: Routledge.

Niaz, M. (2012). *From 'Science in the Making' to understanding the nature of science: an overview for science educators.* New York: Routledge.

Niaz, M. (2016). *Chemistry education and contributions from history and philosophy of science.* Dordrecht: Springer.

Niaz, M., Klassen, S., McMillan, B., & Metz, D. (2010a). Reconstruction of the history of the photoelectric effect and its implications for general physics textbooks. *Science Education, 94,* 903–931.

Niaz, M., Klassen, S., McMillan, B., & Metz, D. (2010b). Leon Cooper's perspective on teaching science: an interview study. *Science & Education, 19,* 39–54.

Pinnick, C. (2005). The failed feminist challenge to 'fundamental epistemology'. *Science & Education, 14,* 103–116.

Polanyi, M. (1964). *Personal knowledge: towards a post-critical philosophy*. Chicago: University of Chicago Press. (first published 1958).

Polanyi, M. (1966). *The tacit dimension*. London: Routledge & Kegan Paul.

Popper, K. R. (1962). Die logic der sozialwissenschaften. In T. W. Adorno et al. (Ed.), *Der positivismusstreit in der deutschen soziologie* (pp. 103–123). Neuwied: Luchterhand.

Posner, G. J., Strike, K. A., Hewson, P. W., & Gertzog, W. A. (1982). Accommodation of a scientific conception: toward a theory of conceptual change. *Science Education, 66*(2), 211–227.

Quine, W. V. O. (1969). Epistemology naturalized. *Ontological relativity and other essays* (pp. 69–90). New York: Columbia University Press.

Restivo, S. (1994). *Science, society and values: towards a sociology of objectivity*. Bethlehem: Lehigh University Press.

Roscoe, K. (2004). Lonegran's theory of cognition, constructivism and science education. *Science & Education, 13*, 541–551.

Rowlands, S., Graham, T., & Berry, J. (2011). Problems with fallibilism as a philosophy of mathematics education. *Science & Education, 20*(7–8), 625–654.

Slezak, P. (1994). Sociology of scientific knowledge and scientific education, Part I. *Science & Education, 3*(3), 265–294.

Smith, M. U., Siegel, H., & McInerney, J. D. (1995). Foundational issues in evolution education. *Science & Education, 4*(1), 23–46.

Smolicz, J. J., & Nunan, E. E. (1975). The philosophical and sociological foundations of science education: the demythologizing of school science. *Studies in Science Education, 2*, 101–143.

Snow, C. P. (1963). *The two cultures and a second look*. New York: Cambridge University Press.

Stuewer, R. H. (1975). *The Compton effect: the turning point in physics*. New York: Science History Publications.

Suppe, F. (1977). *The structure of scientific theories*. Chicago: University of Illinois Press. (2nd edn.).

TEK-NA projektet. (1975). *Syo-konsulentbroschyr*. Stockholm: National Library of Sweden.

Tiberghein, A., Cross, D., & Sensevy, G. (2014). The evolution of classroom physics knowledge in relation to certainty and uncertainty. *Journal of Research in Science Teaching, 51*(7), 930–961.

Vermeir, K. (2013). Scientific research: commodities or commons? *Science & Education, 22*(10), 2485–2510.

Wong, S. L., Kwan, J., Hodson, D., & Jung, B. H. W. (2009). Turning crisis into opportunity: nature of science and scientific inquiry as illustrated in the scientific research on severe acute respiratory syndrome. *Science & Education, 18*(1), 95–118.

Chapter 4
Understanding Objectivity in Research Reported in the *Journal of Research in Science Teaching* (Wiley-Blackwell)

4.1 Method

The *Journal of Research in Science Teaching* (JRST) is the official journal of the US-based *National Association for Research in Science Teaching* (NARST), which has members in many countries around the world. JRST started publishing in 1963 and is indexed in the *Social Sciences Citation Index* (Thomson-Reuter). In February 2016, I made an online literature search on the website of JRST, with the keyword "objectivity." (http://onlinelibrary.wiley.com/journal/10.1002). This gave a total of 120 articles published since 1992. All articles were downloaded and a preliminary examination showed that 10 articles could not be included in the study due to the following reason: In these articles the authors provided a reference and the word "objectivity" appeared in the title of that reference. Finally, a total of 110 articles were evaluated on the same criteria (Levels I–V) as in the previous study (see Chap. 3). Following the guidelines based on Charmaz (2005), presented in Chap. 3, and in order to facilitate credibility, transferability, dependability, and confirmability (cf. Denzin & Lincoln, 2005) of the results, I adopted the following procedure: (a) All the 110 articles from *Journal of Research in Science Teaching* were evaluated and classified in one of the five levels (for levels see Chap. 3); and (b) After a period of approximately 3 months all the articles were evaluated again and there was an agreement of 94% between the first and the second evaluation. It is important to note that all the articles evaluated in this study referred to objectivity in some context, which may not have been the primary or major subject dealt with by the authors. Detailed examples of all five levels are presented in the next section. A complete list of the 110 articles from *JRST* that were evaluated is presented in Appendix 3. Distribution of all the articles according to author's area of research, context of the study and level (classification) is presented in Appendix 4.

© Springer International Publishing AG 2018
M. Niaz, *Evolving Nature of Objectivity in the History of Science and its Implications for Science Education*, Contemporary Trends and Issues in Science Education,
DOI 10.1007/978-3-319-67726-2_4

4.2 Results and Discussion

Each of the 110 articles from *JRST* was evaluated (Levels I–V) with respect to the context in which they referred to objectivity. Based on the treatment of the subject by the authors 21 sections (categories) were developed to report and discuss the results. These sections are presented in alphabetical order. Distribution of the 110 articles according to the Level was the following: Level I = 4; Level II = 33; Level III = 68; Level IV = 5; and Level V = none. It is important to note that some of the articles could have easily been placed in more than one section. The idea behind creating 21 categories (sections) is to facilitate the reader to find the subject of her/his interest. Given the wide range of subjects discussed by the authors over a period of more than 20 years, it is difficult to create the semblance of a continuous storyline (as suggested by one of the reviewers). Similarly, due to limitations of space it is not possible to present a detailed critical analysis of every article. The following are the 21 categories (sections) that were created to present and discuss the results.

4.2.1 Alternative Methodologies and Objectivity

A major responsibility of science education researchers is to generate knowledge and understanding that can influence practice (Yeany, 1991). In this context, either/or dichotomy between qualitative and quantitative methods of research is misleading and even perhaps an obstacle in our endeavors to influence practice. Based on the framework provided by Habermas (1972), Kyle, Abell, Roth, and Gallagher (1992) have argued that following alternative research methodologies can also help to make our knowledge more objective:

> Clearly, the sciences must seek to preserve their objectivity in the face of particular interests. Although such *objectivity is possible to some degree under certain circumstances,* we must acknowledge that fundamental cognitive interests can influence the very objectivity we seek to preserve (p. 1016, italics added). Classified as Level III.

In a similar context, Niaz (1997) has shown that competition between alternative research methodologies (qualitative and quantitative) in science education can provide a better forum for a productive sharing of research experiences. Recent research has shown the importance of mixed methods research programs (Tashakkori & Teddlie, 2003), that at the same time facilitate competition between divergent approaches to research in science education (for a rationale based on history and philosophy of science, see Niaz, 2011, Chap. 3).

In order to facilitate Australian secondary school students' understanding of genetics, Tsui and Treagust (2007) developed a multidimensional conceptual change framework based on an interpretive approach that used multiple data collection methods. Findings suggested that multiple representations facilitated conceptual change.

With respect to the qualitative research rigor of their methodology, the authors concluded:

> In keeping with the interpretive research paradigm, we used, as Guba and Lincoln (1989) suggested, credibility/transferability, dependability, and confirmability in place of internal/ external validity, reliability, and objectivity, which experimental research uses .... The analysis and interpretation of data generated explanations that led to formulation of assertions to be confirmed or disconfirmed through triangulations (e.g., data, methodological, and theoretical triangulation) .... Such research strategies were used to improve the quality and credibility of the data collected in this study, address the research limitations, and thus increase the rigor of qualitative research (p. 212). Classified as Level III.

This presentation implicitly recognizes the problematic nature of objectivity in experimental research and hence the need to complement with an alternative methodology, namely qualitative research. Furthermore, triangulation facilitated not only the rigor of the research findings but also helped to understand its limitations.

Venville (2004) designed a study based on qualitative data collection methods to investigate the process of conceptual change in young Australian children (5- and 6-year old) while they were engaged in learning about living things. The social milieu of the classroom context exposed students' scientific and nonscientific beliefs that facilitated conceptual change. Venville (2004) attributed this change primarily to the methodological aspects (trustworthiness) employed in the study:

> The social constructivist-based research drew on Guba and Lincoln's (1989) notion of trustworthiness to ensure overall quality rather than the traditional standards of rigor in positivistic styles of research (Lincoln & Guba, 2000). Traditional terms of internal and external validity, reliability, and objectivity are replaced by notions of credibility, transferability, dependability, and confirmability .... The credibility of the research findings in this study was enhanced by the use of triangulation so that two or more sources of data, data-collection techniques, ... were employed (p. 460). Classified as Level III.

An important feature of this presentation is that it replaces the traditional criteria of research associated with objectivity, with more qualitative criteria based on triangulation. A major premise of this methodological innovation is to make research more meaningful for classroom practice. Furthermore, this study explicitly associates the traditional quantitative methodological framework with positivistic styles of research, and thus shows the need for "transgression of objectivity" (see Chap. 1).

Recent research has recognized the importance of augmenting students' formal school science experiences with informal experiences outside of the classroom that facilitate an intersection of students' personal knowledge with canonical disciplinary knowledge. One of the informal educational settings is provided by robotics and robotics competitions. Verma, Puvirajah, and Webb (2015) designed a study to investigate high school students' linguistic and social activities in a regional (USA) robotics competition as a sociocultural activity in an informal setting. Based on Critical Discourse Analysis, this study offers conceptual insights into how the culture of the robotic activities is constructed by the students and their mentors. Authors highlight the following methodological aspect of their study based on an interpretive framework where realities are constructed in the human mind based on social and environmental interactions:

> Thus, the intent of our approach is to understand and interpret the contexts or the ecological conditions under which our participants carry out their social activity during the competition. As such, the contextual interactions between our participants and us (as researchers) allow for the creation of certain realities that personify "the importance of subjective human creation of meaning, but does not reject outright some notion of objectivity" (Crabtree & Miller, 1999; p. 10). (p. 273). Classified as Level III.

This presentation explicitly posits the role played by subjective human creation in the process of meaning making. At the same time it also recognizes the inevitable relationship between objectivity and the underlying subjectivity that leads to the creation of multiple realities that may exist as a result of socially constructed and subjective interpretations of meaning making systems. In a sense this represents a good example of how Daston and Galison (1992) understand this tension in understanding the difference between subjectivity and objectivity: "Objectivity is related to subjectivity as wax to seal, as hollow imprint to the bolder and more solid features of subjectivity. Each of the several components of objectivity opposes a distinct form of subjectivity; each is defined by censuring some (by no means all) aspects of the personal" (p. 82). Indeed, every personal construction of the students (subjective) can always be contrasted with the objective canonical knowledge, which thus leads to a new reality.

4.2.2 Assessment and Objectivity

Briscoe (1993) has recounted the personal struggles of a high school chemistry teacher (Brad) who primarily used multiple-choice instruments for evaluating students, which facilitated assessment as there was always one right answer to a given question. However, this perspective started to change as Brad adapted the curriculum to be consistent with a problem-centered approach to learning. While designing alternative means of assessment, Brad recognized that students deserve to get credit for their personal sense-making of chemistry content. Nevertheless, he was at the same time concerned that he may not be consistent and might allow his personal feelings about a student to interfere with his evaluation. He expressed this in the following terms: "If somebody came up with a response that in any way, had some reasonable chemistry in it in relation to what the question was, right or wrong, they got at least a 1 ... I sit down and try to think through an answer that seems reasonable to me. And then I start looking at some of theirs and read over a bunch of them and get an idea of what they have produced and then try to come to some balance between the two" (Reproduced in Briscoe, 1993, p. 981). Over time, Brad experienced a "cognitive disequilibrium" as he felt that his assessment strategy did not comply with the accepted norms of his peers, formal training as a chemistry teacher and that multiple-choice assessments were the accepted norm for evaluating students. Finally, "Brad conceptualized that in assessment contexts, personal interpretations by teachers of students' responses in order to make decisions as to their viability introduced the possibility that the teachers' subjectivity would bias marking decisions. Brad viewed *subjectivity* as bad and *objectivity*, not allowing oneself to be influenced by personal interpretations, as good" (Briscoe, 1993, p. 980, italics added). Classified as Level III.

I am sure many teachers in different parts of the world must have experienced the same dilemma as that faced by Brad. Assessment is a complex issue and has many facets. First, the research community itself has recognized that multiple-choice questions are generally algorithmic and do not facilitate conceptual understanding (Nurrenbern & Pickering, 1987). In a similar vein, Niaz and Robinson (1993) have shown that the ability to solve computational problems (based on algorithmic solution strategies) is not the major factor in predicting success in solving problems that require conceptual understanding. Second, it would be interesting to compare algorithmic problems (found in most textbooks) and conceptual problems that do not necessarily use the multiple-choice format. Consider the following question (multiple-choice) related to atomic structure:

In Rutherford's alpha particle scattering experiments, which of the following statements is correct:

(a) Almost all the alpha particles passed through the thin metal foil undeflected (correct response).
(b) A great number of alpha particles were deflected at large angles.
(c) Almost all the alpha particles bounced back toward the particle source.
(d) None of the above.

Such problems can be found in many science textbooks in different parts of the world and would perhaps be considered as part of "objective assessment." In contrast, let us consider the following question (again based on Rutherford's experiments):

How would you have interpreted, if most of the alpha particles would have deflected through large angles?

This question formed part of a study conducted by Niaz, Aguilera, Maza, and Liendo (2002), in order to facilitate freshman students' conceptual understanding. It is important to note that the experimental data in this question is not the same as found by Rutherford. On the contrary, students are being asked to respond to a hypothetical situation, in which "most of the alpha particles would have deflected through large angles." This change made the problem relatively difficult and also novel as compared to the previous question presented in the multiple-choice format. Based on the experimental treatment used in this study, 20% of the students provided conceptual responses and following are five examples (reproduced in Niaz et al., 2002, p. 518):

1. That the atom did not consist of empty space, but rather the nucleus was about as big as the atom itself. As the nucleus was charged positively, most of the alpha particles deflected through large angles.
2. The nucleus then must have been bigger than what Rutherford had proposed. Consequently, the electrons could not rotate around the nucleus and perhaps it

looked more like Thomson's model. It also suggests that the atom for the most part was not empty ....

3. If most of the alpha particles would have deflected through large angles ... this would have suggested that the model proposed by Thomson coincided more with reality ....

4. That the nucleus is sufficiently big, so as to impede the passage of most alpha particles ....

5. Rutherford's experiment (a small number of alpha particles deflected through large angles) led him to propose the concentration of the positive charge in a small part (nucleus) of the atom. Now, if the contrary had happened, that is, most of the alpha particles would have deflected through large angles, Rutherford would have arrived at a different conclusion. He would have deduced that the atom consisted of a nucleus considerably greater, positively charged, and occupied most part of the atom ....

Responses #2 and #3 are particularly interesting as they refer to Thomson's model of the atom which postulated that the positive charges were uniformly distributed throughout the atom. Actually, after Rutherford (1911) proposed his nuclear model of the atom, a bitter controversy ensued between Thomson and Rutherford that lasted for many years (cf. Niaz, 1998, 2009; Wilson, 1983). This controversy was discussed in class as part of this study (Niaz et al., 2002), and a reference to Thomson's model in this question by two of the students shows how students can incorporate new information creatively. Now, let us go back to the dilemma faced by Brad and compare the possible responses in the multiple-choice question and the new format of conceptual understanding. Even if we accept Brad's qualms with respect to subjectivity–objectivity in assessment, the issues involved go far beyond and provide an opportunity to reflect upon the very essence of the scientific enterprise, namely doing and understanding science involves interpretation and not a simple regurgitation of experimental details (quite similar to the multiple-choice question format presented above). This clearly shows the need for including conceptual problems in the science curriculum and textbooks that can provide teachers an opportunity to assess knowledge more meaningfully, and at the same time facilitate an understanding of the scientific enterprise. A recent study has called attention to the need for emphasizing conceptual understanding: "If we turn however to matters of conceptual understanding, we realize that our students are as a rule ignorant and cannot answer questions such as: why chlorine appears with so many oxidation numbers, why spontaneous endothermic reactions exist, and why reactions lead in general to chemical equilibrium" (Tsaparlis, 2014, p. 42).

Higher-order cognitive skills (HOCS) refers to activities such as question asking, problem solving, decision-making, critical, and evaluative thinking. In contrast, problems based on lower-order cognitive skills (LOCS) require simple recall information or an application of algorithmic processes to familiar situations and contexts. HOCS problems are generally unfamiliar to the students and require application of known theories to unfamiliar situations. Students are generally concerned about those problems that they consider important from the perspective of the final assessment (exam) and the traditional educational structure reinforces such expectations. Based on these considerations, Zoller (1999) has concluded that the development of students' HOCS requires appropriately designed HOCS-type examinations which:

> ... constitute an antithesis to the existing dominant objective-type exams. However, the difficulties and time limitations associated with the design, administration and grading of HOCS-oriented exams constitute a barrier for their implementation. Therefore, attempts in this direction at universities with large lecture sections may be rejected by faculty claiming that this kind of examination is unmanageable timewise (as far as evaluation and grading are concerned) and that the *objectivity* in students' grading cannot be guaranteed. Quite often, the objection(s) to HOCS-type examinations are supported by university authorities either on the philosophical-ideological or pragmatical levels, and thus may adversely affect the desired LOCS to HOCS shift in chemistry and science teaching (pp. 585–586, italics added). Classified as Level III.

It is important to note that assessments based on LOCS-type exams are generally multiple-choice items, corrected by the computer and apparently most teachers consider them to provide objective evaluation of students' achievement. In such evaluations chemistry (also other areas of science) knowledge is considered to be a rigid body of facts revealed by authority (professor or text) and students simply respond without interacting with the content. It is plausible to suggest that under such circumstances it is difficult for the students to develop problem-solving and decision-making capacities that are important for a responsible citizenry. So the issue is not of facilitating objectivity in evaluation but rather depriving students of an environment that can be more meaningful and rewarding in the long run.

Based on the work of critical, feminist, and multicultural science educators, Fusco and Barton (2001) have developed a perspective that they refer to as critical science education. This enabled them to develop a conceptual argument for expanding current visions of performance assessment to include: value-laden decisions about what and whose science is learned and assessed and include multiple world-views. Furthermore, assessment is a method and an ongoing search for method. Based on Gipps (1999) they argue that assessment is value laden, socially constructed and that it is not an exact science. Based on their experience in a youth-led community science project in the inner city (New York), they concluded:

> Teaching science cannot be reduced to the acquisition or mastery of skills or techniques but must be defined within a discourse of human agency. The teaching of science occurs within the larger contexts of culture, community, power, and knowledge. Science teaching therefore must respond to the political and ethical consequences that science has in the world, and must be equally infused with analysis and critique as it is with production, refusing to hide behind *modernist claims of objectivity and universal knowledge*. Teachers help to construct the dynamics of social power through the experiences they organize and provoke in classrooms (Fusco & Barton, 2001, pp. 342–343, italics added). Classified as Level III.

Indeed, assessment is crucial for the educational enterprise, especially if we want the students to be motivated and be a part of the classroom dynamics. Furthermore, even if we accept the traditional assessment methods such as multiple-choice tests, there is no way to know if we are facilitating conceptual understanding. Consequently, it is plausible to suggest that assessment needs to be an ongoing search for method.

Lynch (1994) has explored the policy trends with respect to ability grouping in K-12 science education in the USA. After reviewing the policies of various public and academic agencies the author concluded:

> Specifically, if an educational practice has the result of creating "racially identifiable" classes, then the public educational agency responsible must be able to *defend the objectivity of its grouping practice*, show that improvement in achievement has occurred as a result of the grouping practice, and demonstrate that its grouping practice is more successful than equally effective alternative grouping practices that result in less racial disproportionality (p. 112, italics added). Classified as Level II.

Despite recognizing the role of objectivity as an academic objective (Level II), Lynch (1994) also recognized that ability grouping is a controversial and value-laden topic and that grouping practices alone are unlikely to influence science education reform that requires comprehensive restructuring at the local school level (p. 105).

Roth and McGinn (1998) consider that grades are representational artifacts that are inherently political in that they embody the ideologies and agendas of their designers. Furthermore they found that depending on the grades the relationship between students and teachers was construed differently. For example, students with high grades generally approved of the learning environment, whereas those with intermediate and low grades wanted change but did not voice their needs for fear of repercussion. Drawing on the work of Foucault, Giroux and Latour, these authors concluded:

> In the past, schools have consistently denied and thereby DELETEd the relevance to the learning process of students' "memories, families, religions, feelings, languages, and cultures that give them a distinctive voice" (Giroux, 1992, p. 17). We see in a critical investigation of grading practices one aspect of interrupting representational practices more generally; therefore educators can interrupt representational practices that make claims to universality, objectivity, and consensus, practices that marginalize and DELETE diverse student cultures and their histories in terms of gender, race, and socioeconomic status (Roth & McGinn, 1998, p. 416). Classified as Level III.

Indeed, in most educational systems grading practices are far from being objective and generally do represent the epistemological orientations of the culture, history, and the society.

Reform efforts have emphasized the importance of rigorous standards backed by quality curricula and effective teaching in order to achieve high levels of success in science for all students (AAAS, 1993; NRC, 1996). However, it is not clear how the achievement gap separating low-income, linguistic, racial, and ethnic minority students from more economically privileged students will be accomplished or at least diminished. Warren, Ballenger, Ogonowski, Rosebery, and Hudicourt-Barnes (2001) explored this problem in the context of two case studies based on Haitian American and Latino (5th and 6th grade) students. A basic premise of the study was based on the following rationale:

> The term *scientific* is commonly used to denote a sphere of human activity characterized by special qualities: rationality, precision, formality, detachment, and objectivity. This view is

broadly held in society at large, in schools, and even by some scientists themselves. The term *everyday* is commonly used to denote another, opposing set of qualities: improvisation, ambiguity, informality, engagement, and subjectivity. The presumed differences between scientific and everyday activity are often framed as sets of dichotomies, with the left-hand term in the pair being the privileged, scientific one, the one seen as representing a cognitive ideal: precise versus imprecise language, logical versus analogical reasoning, skepticism versus respect for authority, and so forth (p. 530). Classified as Level III.

A close look at the history of science will show that the characteristics of the two terms, scientific and everyday can be used interchangeably by the scientists and even perhaps most students (cf. Niaz, 2009). The possibility of continuous change within these sets of dichotomies can be helpful in teaching science especially to marginalized children. A major strength of this study is that the authors conceptualized children's diverse *everyday* sense-making and *scientific* sense-making as potentially complementary and continuous processes. The findings of this study showed that although in some respects the privileged and the marginalized children differed in their classroom interactions (scientific sense-making), the latter group was equally capable of learning academic knowledge and practice.

4.2.3 Capitalism, Critical Pedagogy, and Objectivity

The relationships among capitalism, science and education have been explored by Barton (2001a) in an interview with Peter McLaren, a Marxist, and a professor in the Division of Urban Education, University of California, Los Angeles. McLaren clarifies that his central claim is that it is not possible to divorce educational policy from the transmogrification of the world economy because the global financial system is overrun by speculators and modern-day robber barons who are concerned with profit at any cost rather than social justice (p. 850). At one stage during the interview, Barton (2001a) expressed her views with respect to science as a culture and practice and that the challenges in urban science education are layered, and these layers are deeply connected to each other and to issues of power and control:

> I am concerned that science education has not incorporated the needs or concerns of children in poverty and children from ethnic, racial, and linguistic minority backgrounds. These "gaps" can be seen in high-stakes tests, mandatedcurricula, and daily school practices. I am also concerned that science—as a culture and practice—has developed along elitist lines resulting in a knowledge base and a cultural practice reflective of those already in power and uses *the unobtainable ideals of truth and objectivity* to hide its singular focus. Finally, I am concerned that schooling itself and the workaday practices of low-level worksheets, discipline through humiliation, and teacher-student bargaining (to name only a few) in urban centers strips children of their cultural identities, their right to learn, and their dignity as human beings (p. 852, italics added). Classified as Level III.

This clearly represents the state of science education in many parts of the world, in which students and perhaps also teachers do not experience an atmosphere that is congenial to creative learning. Again, a basic premise of most educational practice fosters the ideals of "truth" and "objectivity" that are at best not attainable. Continuing with the interview, McLaren emphasizes classroom practice and highlights the work of Sandra Harding (1998) on *standpoint epistemology* which highlights the relationship between knowledge and politics and explains the effects that different kinds of political arrangements have on the production of knowledge and knowledge systems:

> *Empiricism tries to "purify" science. Yet Harding has shown that these empirical methods never reach greater objectivity, for they exclude thought from the lives of the marginalized.* For Harding, who draws upon postcolonial, feminist, and post-Kuhnian social studies of science and technology as well as Latour's notion of technoscience, with its tension between local and global science practices—all attempts to produce knowledge of any kind are socially situated, and some of these objective social locations are better than others as starting points of research. Harding points out that, for instance, when physics is permitted to set the standards for what counts as nature and what counts as science, knowledge becomes truncated and is often misapplied, limiting our ability to produce knowledge in ways that can assist aggrieved populations (p. 856, italics added).

Most critics would agree that practice in science education in many parts of the world is based on an empiricist epistemology that presents to the students a "purified" version of how science develops and progresses. Holton (1969) has referred to this practice as the "myth of experimenticism." The path from experiment to theory or the emphasis on empirical methods not only do not provide greater "objectivity" but also deprive students of an environment that facilitates thinking. The reference to physics as a standard for understanding science and nature has also been questioned in recent philosophy of science.

As the interview continued, Barton pointed out that some critics consider Harding's standpoint epistemology as relativist and it denies to the students the opportunity to "learn the canon" or to "have access to the culture of power" (p. 857), which may further oppress the marginalized community. McLaren responded that "learning the canon" and learning how culture and science intersect are not mutually exclusive (p. 857). Furthermore, Harding does not assume that because a standpoint is articulated from the position of the oppressed that is necessarily the best position.

Recent developments in science and its relationship to the nation, state, and private commercial interests have been referred to as globalization. Carter (2008) has explored the implications of this changing form of science for science education and concluded: "Considering the engagement of science and globalism requires an acknowledgment of the long relationship between science, capitalism, and the world system … [the] argument that science's official story of 'objectivity' and 'autonomy' attempts to diminish the link between science, capital, and market forces, preferring instead the romantic principle of a value-free science" (p. 621, Classified as Level III). The argument for a value-free science is difficult to sustain as most human activities are value-laden, and historians of science have recognized this facet of the progress in science (cf. Machamer & Wolters, 2004). Furthermore, the relationship between science and Western industrial capitalism is well established, "While I reject some of the components of the notion that *science is value free* … I too think that there is a significant distinction between cognitive and social values. I will argue that the distinction is crucial for properly interpreting the results of scientific research and for opening up reflection on how neutrality might be defended as a value of scientific practices at a time when much of mainstream scientific research is becoming increasingly subordinate to 'global' capitalism" (Lacey, 2004, p. 25, original italics). This clearly shows that although historians and philosophers of science (also science educators) would aspire for a science that is value free, neutral and even objective, the real picture of the scientific enterprise is too complex to follow such simple schemata.

4.2.4 Constructivism and Objectivity

Positivist views of science focus on the "objective" study of phenomena that emphasize observation and neglect students' previous ideas or beliefs. The existence of objective truths that are domain-specific and constant leads to a structure of knowledge that facilitates rote learning, which on surface seems to be more efficient for learning science. This view contrasts with the constructivist approach based on previous knowledge that evolves continually (Kuhn, 1962). This background provides Edmondson and Novak (1993) to endorse von Glaserfeld's "radical" constructivism, in which truth is based on coherence with our previous knowledge and not on correspondence between knowledge and objective reality. Next these authors refer to a study from their research group in which women scientists were interviewed about their learning, teaching experiences, and research programs, to conclude:

> They are teaching science that they *also* know from their research is not only the product of the process of careful and consistent method, but also the product of the influence of world view, beliefs and changing theory. There is an unreconciled conflict about the apparent dichotomy of the objectivity/subjectivity dualism in the process of science, and the evolving nature of knowledge (Reproduced in Edmondson & Novak, 1993, p. 550). Classified as Level IV.

This statement clearly establishes a relationship between the production of knowledge and world views. The dualism between objectivity and subjectivity leads to a conflict in the evolving nature of progress in science. The reference to the evolving nature of knowledge in the context of science as a process helps to understand the objectivity–subjectivity dualism within a historical context. Ignoring this duality may lead to the hegemony of objective knowledge and the consequent emphasis on rote learning.

While criticizing Piaget's theory of cognitive development, O'Loughlin (1992) has emphasized the role of sociocultural and contextual factors, and how this makes the implementation of constructivism in classroom difficult as at the level of formal operations, the highest stage of reasoning in Piaget's scheme, all content is excluded and the entire reasoning process is described in terms of a set of logical operations:

> The focus on scientific rationality, the interest in describing intellectual advancement in terms of increasing decentration from subjectivity and toward objectivity, and the desire to express the highest forms of reasoning in terms of content-free logical operations all

point to a model of cognitive development in which reasoning that is ahistorical, value-free, and abstract is regarded as the *telos* of cognitive development. From Piaget's perspective the absence of interest in sociocultural and contextual factors can be explained in terms of his exclusive interest in isolating universals of cognitive development. Real difficulties arise, however, when constructivists appropriate this universalist theory to deal with classroom learning processes that are inherently constrained by sociocultural and contextual factors (p. 795, original italics, underline added). Classified as Level III.

O'Loughlin seems to be suggesting that decentration involves a move that goes from the state of subjectivity to objectivity, or in other words finally the state of objectivity is achieved. On the contrary, Piaget (1971) has pointed out explicitly the problematic nature of objectivity:

> ... *objectivity is a process and not a state*. This amounts to saying that there is no such thing as an immediate intuition touching the object in any valid manner but that objectivity presupposes a chain reaction of successive approximations which may never be completed (p. 64, italics added).

This clearly shows that Piaget conceptualized the evolving nature of objectivity to that approximates to the historical perspective presented by Daston and Galison (2007). Furthermore, O'Loughlin (1992) ignores a fundamental distinction in Piaget's *oeuvre*, namely the epistemic and psychological subjects. Although, O'Loughlin does mention these subjects (see pages 794 & 805), it lacks the fundamental importance of this distinction in Piaget's theory of cognitive development. To put it in a historical perspective, Piaget builds a general model by not emphasizing individual differences (however, recognizes their role and importance), that is studies the epistemic subject whereas Pascual-Leone, by incorporating a framework for individual difference variables, studies the metasubject, that is, the psychological organization of the epistemic subject, which is an attempt at explaining performance or specifying process criteria. Pascual-Leone considers his theory of constructive operators to be a, "... model of the psychological organism (*the metasubject*) which is at work inside Piaget's 'epistemic subject' for each age group as much as inside the particular children which educators encounter" (Pascual-Leone, Goodman, Ammon, & Subelman, 1978, p. 271). Similarly, Kitchener (1986) and Niaz (1991) consider that individual difference variables are outside the purview of a "developmental explanation." After having established the difference between the epistemic and psychological subject, it is important to note that Piaget himself recognized this distinction explicitly (although his work primarily dealt with the epistemic subject):

> A fundamental epistemological distinction must be introduced between two kinds of subjects or between two levels of depth in any subject. There is the "psychological subject," centered in the conscious ego whose functional role is incontestable, but which is not the origin of any structure of general knowledge; but there is also the "epistemic subject" or that which is common to all subjects at the same level of development, whose cognitive structures drive from the most general mechanisms of the co-ordination of actions (Beth & Piaget, 1966, p. 308).

Fosnot (1993) has argued that in the 10 years before his death Piaget (1977) revised his model of equilibration by emphasizing that knowledge proceeds neither solely from the experience of objects nor from innate programming performed in the subject but from successive constructions. This leads Fosnot (1993) to conclude that in Piaget's model of equilibration:

> ... both structure and content are constructed [thus] it seems erroneous to ... conclude ... that Piaget does not consider human subjectivity or the social context. In my mind, that is at the heart of his theory—and at the heart of constructivism. There is no such thing as objective thought, because thought is the result of the act of a subjective knower within a social context transforming, organizing, and interpreting with structures previously constructed, but open to accommodation—a dialectical interaction (p. 1193). Classified as Level III.

Fosnot clearly recognizes the role played by subjectivity in the educational context. Next she considers the cognitive constructivists (e.g., Piaget) to be following the internalist program, whereas the social constructivists follow the externalist program. In order to understand this internalist/externalist account in the history of science she draws on Harding's (1987) work, namely the internalist program analyzed the rise of modern science as an endogenous development of intellectual structures, whereas the externalist program emphasized the economic, social, and cultural changes. Again, endorsing Harding's position, Fosnot suggests that science educators need to hear the story that each side tells, that is collaboration between cognitive and social constructivists. In a similar vein Niaz (2011, Chap. 11) has drawn an analogy between the progress in atomic structure (science) and educational practice (constructivism). History of atomic structure in the twentieth century is based on a series of atomic models that developed by including some aspect of the earlier models (e.g., Thomson, Rutherford, Bohr, and wave mechanical models of the atom). Similarly, constructivism has evolved through a sequence based on the work of various scholars (e.g., Piaget, Ausubel, von Glasersfeld, Vygotsky, and Perkins). Continuing this line of reasoning, Niaz et al. (2003) have argued that constructivism in science education (like any scientific theory) will continue to progress and evolve through continued critical appraisals (based on various aspects of the different forms of constructivism). This clearly helps to understand the evolving nature of objectivity through constructivist theory in the educational context.

Based on a questionnaire, Hashweh (1996, p. 49) classified science teachers (Palestine) in the following two groups: (a) Constructivists. These teachers believed that the aim of science was to develop theories to understand the world, absolute *objectivity* was impossible (observations are theory-laden), testing theories against experience was important than their origins, scientific knowledge was tentative and emphasized the importance of scientific revolutions and conceptual change; (b) Empiricists. These teachers believed that the aim of science was to collect facts about the world, scientific knowledge was objective, permanent, and discovered (rather than invented), and emphasized the role of observations, the scientific method, and the gradual and the accumulative aspects of the growth of scientific knowledge. A year later these science teachers responded to another questionnaire and found that teachers holding constructivist beliefs are (p. 47): (1) More likely to detect student alternative conceptions; (2) Have a richer repertoire of teaching strategies; (3) Use potentially more effective teaching strategies for inducing student conceptual change; and (4) Report more frequent use of effective teaching strategies. These results clearly show the advantage of constructivist teaching strategies and at the same time lead to yet another question: Can these epistemological beliefs of the teachers influence and facilitate students' conceptual change? Classified as Level III.

Ritchie, Tobin, and Hook (1997) designed a teaching strategy to facilitate Grade 8 science students' learning of electric circuits in Florida, USA. A basic premise of the study was that from a constructivist perspective, learners construct viable knowledge rather than representations of truth. An important feature of the study was that the researchers observed while the teacher was involved in his classroom activities. This enabled the researchers to provide feedback (and exchange ideas) to the teacher during interviews after the class. These conversations made the teacher more sensitive with respect to the students' alternative conceptions. Authors (the teacher was one of the authors) also recognized that these interactions added pressure and escalated the teacher's frustrations at times. However, the teacher appreciated this involvement and considered that it was good for him and also the kids. Based on this experience, Ritchie, Tobin, and Hook (1997) concluded:

> This raises the old but important issues of objectivity and researcher independence. It is obvious that by affecting Mr. Hook's learning environment as we did, we were not concerned with maintaining a distance between ourselves and the subjects of our investigation. Instead, our role was more supportive so that we could establish and sustain a responsive, mutually acceptable dialogue about classroom events, *audit the process rather than the product*, create a situation in which the teacher was able to reflect systematically on practice, and act as a resource which the teacher could use (p. 236, italics added). Classified as Level III.

Authors' concern for objectivity and researcher independence provides an interesting backdrop to the bigger challenge: "audit the process rather than the product." Indeed, despite some reservations most researchers would accept this as an innovative strategy. Furthermore, this addresses yet another issue with respect to the need for teachers to understand that they must consider themselves also as learners and that their constructions of knowledge are never complete.

4.2.5 Controversy in Science and Objectivity

Cross and Price (1996) have explored the perceptions of teachers with respect to teaching of controversial issues and the possible tension with value-free science curricula. The study is based on secondary school science teachers in Scotland and Connecticut (USA). The role of controversies is difficult to understand as in most parts of the world science is still largely taught as if it were objective and value free and theories are taught as never changing facts. Authors reported that conversations with the teachers turned to the tension due to, "… science teachers' allegiance to value-free objectivity of science and the more modern view of the problematic nature of the scientific knowledge and the interests held by various scientists and stakeholders" (p. 325). Findings of the study led the authors to suggest that learning the concepts of science within a framework of teaching controversial issues is a contentious issue, however, it is not an insurmountable obstacle (p. 325). Classified as Level III.

4.2.6   Critical Ethnography and Objectivity

Barton (2001b) has drawn attention to the need for recognizing that praxis implies theory into action, and this leads to the next stage with respect to what actions are adequate, responsive, and necessary. Drawing on the framework of Freire (1971), she suggests that the most important outcome of research should be the conscientization of both the researcher and participants. She goes beyond and endorses the breaking down of the separation between research and the struggle for social change. Furthermore, this perspective leads to an understanding that justice must be viewed as a more essential measure of the strength of research than its objectivity (p. 911). Critical ethnography can be particularly helpful for such an understanding as it is a kind of methodology that emerges collaboratively from the lives of the researcher and the researched, leading to a praxis committed to the defense of human rights (p. 899). Classified as Level III.

4.2.7   Critical Feminism and Objectivity

Drawing on the work of critical and feminist scholars in science and education, Barton (1998) has emphasized the need for inclusion in science education of urban homeless children. The typical science curriculum ignores the struggles waged by these children to find a place for their lives in science and for science in their lives. To make science available to all involves overcoming situations created by conflicting paradigms:

> Making the pedagogical questions of identity and representation central to the struggle to create a science for all pushes against the historically accepted *modernist frameworks of positivism*, instrumental reason, universal knowledge, and bureaucratic control that have been at the center of curriculum and practice in science education. From the feminist perspective brought to bear on the experiences of homeless children in this article, I have tried to argue that science education can no longer hide behind the *modernist claim to objectivity* and universal knowledge (Barton, 1998, p. 391, italics added). Classified as Level III.

It seems that positivism and its claims to objectivity leads to a conflicting situation that does not facilitate the inclusion of all in science. The relationship between logical positivism and mechanical objectivity has been recognized by Daston and Galison (2007).

Following their suffragist foremothers in the early twentieth century Michigan (USA), Cavazos et al. (1998) have issued a *call to action*, in order to bring attention to the marginalized voices of feminists in the struggle to implement "science for all." The central question of this call to action is: what implications do feminism, critical theory, and post-structuralism have for science teaching and research? These authors first set out to outline what people expect them to do and think (p. 342): (a) Researchers should maintain an objective voice and present their findings in standard modes of scientific reporting; (b) Research should be dispassionate and devoid of emotional character to avoid bias in findings and interpretations; and (c) There is no room for personal voice in scholarly writing. Indeed, this is what most researchers in principle do aspire to do and perhaps achieve. Interestingly, it would be interesting to inquire if such guidelines are actually implemented. Even methodology courses and textbooks would endorse such an agenda. However, anyone who has done research work in science or science education knows that at best this agenda represents a chimera. After outlining the agenda, Cavazos et al. (1998) respond in the following terms:

> *This reporting style masks the subjectivity of all science in a false guise of objectivity.* Research that matters is motivated by deep commitments and passions to learn. Feminism insists that we acknowledge these passions and the emotional as well as intellectual lives of researchers. Researchers should make visible their personal biases, values, and commitments in reporting their research (p. 342, italics added). Classified as Level III.

History of science shows that, science in the making is characterized by the work of researchers who are deeply committed to their passion to learn. Interpretation of data and events always has an element of bias and even perhaps prejudice. Building consensus in science is a complex social process of competitive cross-validation by the peers (cf. Campbell, 1988a, b).

Bianchini, Cavazos, and Helms (2000) have explored practicing scientists and science teachers' beliefs and experiences related to issues of gender and ethnicity in science education and how to address such issues in science classrooms. A basic premise of the study is that feminist scholars of science question the conventional definitions of what counts as science and how science works: "… they question science's claims of objectivity, value neutrality, universality, and epistemic privilege …. Helen Longino (1990), for example, dismissed the notion of science as a value-free enterprise. For science to become less oppressive, she argued, scientists must deliberately embrace political commitments and explicitly recognize these commitments when making decisions about the truth of knowledge claims" (p. 515). Based on interviews with science teachers and scientists, these researchers elaborated the following continuum to understand nature of science:

Science as objective and universal → Particular aspects influenced by gender and culture →Science as embedded in social, political, and cultural contexts.

These authors relate the experience of a chemistry teacher (Maria) of mixed ethnicity who as a student had a discussion with her English teacher with respect to the interpretation of a poem. The teacher considered Maria's interpretation as "not right," which made her think that English is subjective and thus inclined toward chemistry which facilitated the control of some experimental variables and thus more objective. This shows that classroom experiences provide the environment that can facilitate the formation of a particular understanding of the *two poles (objective– subjective) of the continuum*. It also shows the need to study and understand chemistry (also science in general) within a history and philosophy of science perspective, which reveals that science in the making is replete with controversies and alternative interpretations (for details see Niaz, 2016). Finally, Bianchini, Cavazos, and Helms (2000) concluded: "… examining equity issues in sophisticated ways—balancing recognition of systematic gender and ethnic bias in science with sensitivity to diverse interests and experiences that exist within each underrepresented group—is a necessary but not sufficient step toward achieving an equitable and excellent science education. Equally important is confronting the pervasive, often unconscious assumptions that gender equity is a women's issue and that multiculturalism is a matter of ethnic minorities" (p. 542). Classified as Level III.

4.2.8   Cultural Diversity and Objectivity

Globalization leads science educators to confront the challenge of cultural diversity in their classes. However, given the canonical nature of school science valuing and keeping this diversity is difficult. In order to face this challenge, Van Eijck and Roth (2011) have emphasized the role played by representation in the following terms:

> Representation is the fundamental human characteristic that constitutes the necessary condition for consciousness; representation allows some immanent present to be made present again, to be re/presented … Representation in fact is a necessary condition for the objectivity of science and for its historicity … Understanding the production of culture through representation starts with the recognition that communication occurs by means of signs, which embodies a signifier-signified relation (p. 827). Classified as Level I.

The academic achievement gap between African American and White students in urban science classrooms in the USA is a constant source of challenge for the science education community. Based on Ogbu's (1978) cultural ecological theory, Norman et al. (2001) have explored students' responses to societal disparities. A major premise of the study is that the achievement gap in urban science classrooms reflects the sociocultural position of groups (not racial differences) within society along a spectrum from dominant to marginalized. Authors argue that students who are socioculturally disadvantaged respond to these disparities in ways that impede learning, due to the students' stance of opposition to the school environment and requirements. A functional approach to culture provides a framework for the exploration of student responses and identity formation as manifestations of the interplay between socially oppressive forces and potentially liberating action, and suggest:

> … the potential for oppositionality to become a positive rather than a negative motivation. Culture mediates what behavior means; thus, *from a dominant culture's point of view oppositionality is insubordination, whereas from the minority point of view it is a way of*

preserving identity. We recommend that the science education community recognize more deeply the cultural dimension of science as an intellectual discipline. Patterns of discourse in science may pose challenges for urban students, ranging from simple meaning making in ways at odds with their everyday experience to *canonizing objectivity in a manner that reinforces the privileges enjoyed by society's elite.* (Norman et al., 2001, p. 1111, italics added). Classified as Level III.

Such considerations are an essential component of an adequate appraisal of the institutional and cultural contexts that underlie achievement differences among groups. In this account, *oppositionality* plays a salient role, as for the dominant group it means insubordination, whereas for the minority group it is an attempt to preserve identity. In other words the pattern of discourse of the minority students leads the majority group to consider its understanding as the canonized version of objectivity. The functional approach in contrast can facilitate cooperation (instead of conflict) between divergent perspectives.

4.2.9 Culture of Power and Objectivity

Barton and Yang (2000) have explored the relationship between cultural and socioeconomic issues and the science education of inner-city students. Recent reform efforts (AAAS, 1993; National Research Council, 1996) have emphasized that all children can learn science regardless of age, sex, cultural, or ethnic background. In most parts of the world there is a dominating culture of power that decides how science gets defined and how science is taught and practiced. This culture is not conducive not only toward egalitarian policies but also distorts the very structure of science:

> Classroom activities, such as labs and projects, seldom reflect the "real work" of scientists … For example, most adult scientists spend relatively little time copying facts and definitions out of books, yet that is the primary activity of students in many science classes …. Textbooks and other curricular materials often hide the people, tools, and social contexts involved in the construction of science. The result is often a fact-oriented science which appears decontextualized, objective, rational, and mechanistic (Barton & Yang, 2000, p. 875). Classified as Level III.

Furthermore, these authors stress that science labs and classrooms are typically structured hierarchically with the teacher and the text controlling what aspects of knowledge count. The image of the scientist most frequently projected in science curricula is that of the western self-assured, technologically powerful manipulator and controller (Hodson, 1993). It has even been shown that when children are asked to draw a scientist, the most common drawing is that of a white man wearing a lab-coat and glasses. It is plausible to suggest that the culture of power discriminates more students from underprivileged backgrounds and a democratic society needs to introduce reforms that provide opportunities for inclusion and a science education that goes beyond the rhetoric of the textbooks (cf. Gooday, Lynch, Wilson, & Barsky, 2008).

In the context of teaching the nature of science to elementary science teachers, Bianchini and Colburn (2000) have highlighted the difficulties involved in presenting a cogent and comprehensive picture of what science is and how scientists work. One of the authors (Bianchini) particularly emphasized the contextual aspects of science (personal, social, and cultural values) that are "... deeply enmeshed in the historical, political, cultural, and technological fabric of society, phenomena no different in many respects from other social institutions and cultural practices" (pp. 179–180). Next she relates the findings of the work of Wertheim (1995) who documented how women mathematicians and physicists were systematically denied access to educational opportunities, formal appointments to academic positions, and/or proper recognition for their accomplishments because of their gender. Next she recounts how with the onset of phrenology, scientists started to study human skulls and to use differences in shape and size to argue that black men could be compared to white women, these in turn could then be contrasted with the superior white man. According to Hubbard (1988) as science and technology are enmeshed in politics and power, they "... always operate in somebody's interest and serve someone or some group of people. To the extent that scientists pretend to be neutral they support the existing distribution of interests and power" (p. 13). Bianchini endorsed this position and: "... called for recognition of the political nature and content of scientific work, for the elimination of appeals to objectivity (distancing self from subject) and value neutrality in science" (p. 180). This account clearly shows how in the case of gender and phrenology objectivity and neutrality of the scientific enterprise was compromised and that such epistemic virtues are acquired in degrees and require considerable time and effort (cf. Machamer & Wolters, 2004). Classified as Level III.

The *structure-agency* dialectic is an important tool for framing equity in science education. According to Sewell (1992), *structure* is identified as mutually sustaining cultural schemas and sets of resources that empower and constrain social action and that tend to reproduce by that social action. Structures include school's physical architecture as well as time-based dimensions, such as: duration of the school day, the academic calendar, the written rules, the unspoken norms, and the imposed policies that influence educational processes. On the other hand, *Agency* as the school leaders' capacity to reinterpret and mobilize an array of resources in terms of cultural schemas, as they interpret situations and initiate efforts to alter structures to mitigate challenges. Based on this background, Wenner and Settlage (2015) have explored the leadership practices of school administrators and principals with the following premise: "Activity (e.g., resistence, progress, etc.) emerges via the reflective engagement by an actor [school leader] via enactments of 'subjective' drives against the 'objectivity' of the structural circumstances" (p. 505). Classified as Level III. This comes quite close to what Daston and Galison (1992, 2007) have referred to as a constant struggle between subjectivity and objectivity in the history of science. Based on the structure/agency perspective, Wenner and Settlage (2015) found that principals were found to engage in the cognitive professional practice of *buffering* in four ways: adjusting school structures to accommodate new policies; negotiating compromises with the central office about policy implementation; shielding teachers from low-priority policies; and occasionally encouraging teachers to pre-emptively engage in district-level representation to shape policy implementation. Finally, the authors concluded that principal buffering contributes to the equitability and excellence of student performance on their schools' statewide science test. These findings show clearly the inherent struggle between the two epistemic virtues (subjectivity and objectivity) and how reform movements can go about to introduce changes so that the schools can accomplish their egalitarian objectives.

4.2.10 Feminist Epistemology and Objectivity

Feminist epistemology critiques the traditional conceptions of what counts as knowledge, and the work of feminists such as Evelyn Fox Keller, Donna Haraway, and Sandra Harding has been particularly influential. Like other forms of knowledge it is culturally situated and therefore reflects the gender and racial ideologies of societies. Science cannot produce culture-free, gender-neutral knowledge because Enlightenment epistemology itself is imbued with cultural meanings of gender. According to Brickhouse (2001):

> This feminist critique of Enlightenment epistemology describes how the Enlightenment gave rise to dualisms (e.g., masculine/feminine, culture/nature, objectivity/subjectivity, reason/emotion, mind/body), which are related to the male/female dualism …, in which the former (e.g., masculine) is valued over the latter (e.g., feminine). These dualisms are of particular significance to scholars writing about science because culturally defined values associated with masculinity (i.e., objectivity, reason, mind) are also those values most closely aligned with science (Keller, 1985). As such, not only was masculine culturally defined in opposition to feminine, but scientific was also defined in opposition to feminine (p. 283). Classified as Level III.

These dualisms pose considerable difficulties for changing the present authoritarian structure of most classroom environments. Understanding objectivity itself needs to be situated in a historical context that facilitates recognition of the complexities involved in the production of scientific knowledge. In recent years considerable work has been done on feminist literature related to teaching science and its assessment (some of this literature is reviewed in other parts of this chapter).

Based on feminist epistemology (Harding, Keller, others), Howes (1998) has questioned the "distancing" stance approved by Western scientific objectivity. This is all the more important in teacher education, as teachers' work is embedded in classrooms that involves designing of activities that create expectations that are related to students' success. Furthermore, teachers change their plans in accordance with students' learning on a daily basis as well as over time. With this background the author explored how high school sophomore girls (enrolled in a genetics course in USA) expressed their relationship to and understanding of prenatal testing and its possible place in their lives. It was found that participants used the word "baby" regularly to refer to what is more properly named the "embryo" (3 months since conception) or the "fetus" (3–9 months into pregnancy). Interestingly, the teacher also used the word "baby" on the grounds that use of scientific vocabulary makes learning more difficult and this led Howes (1998) to conclude:

> This situation brings to mind a prevalent pedagogical issue: How and when should we allow unscientific vocabulary to be left unnoted? Lost in the translation from scientific to every-day language are distance, objectivity, and clarity; gained in the translation are intimacy, complexity, and untidiness. If it is true that we can help students connect with science by avoiding alienating scientific knowledge, what then? Should I have insisted that the students say "fetus" or "embryo"? Or did allowing them—even encouraging them, by my own language—to use the words of their choice allow them access to knowledge that may have been denied them if other words were used in our discussions? I hoped to help them be intrigued throughout the unit, and thus hesitated to impose more scientific vocabulary on them (p. 891). Classified as Level III.

Finally, although the overuse of scientific vocabulary may succeed in confusing and alienating more students than it intrigues, Howes also recognized that as a feminist (prochoice) she was concerned that recognizing the fetus as human by calling it "baby" may have influenced these girl students to believe that abortion is murder.

Brotman and Moore (2008) have reviewed literature on gender and science education and found four themes: equity and access, curriculum and pedagogy, the nature and culture of science, and identity. While focusing on the nature and culture of science they concluded:

> One more important idea underlying many of the studies in this theme is the commonly made link between masculinity and traits such as objectivity, rationality, and lack of emotion, which are also often associated with science. This *association between masculinity, objectivity, and science* does two things. First, because femininity is viewed as mutually exclusive with masculinity, femininity also becomes viewed as mutually exclusive with science .... Second, science becomes viewed as unassociated with traits culturally defined as feminine, such as subjectivity, emotion, and creativity (p. 987, italics added). Classified as Level III.

Indeed, associating objectivity with masculinity ignores an important facet of the progress in science in which there has been a constant struggle between the notions of subjectivity and objectivity (cf. Daston & Galison, 2007). Consequently, following the historical evolution of objectivity in the history of science can also facilitate a better understanding of gender in science education.

Hildebrand (1998) has argued that the hegemonic writing practices typified by science laboratory reports compound the difficulties due to their heavy reliance on objectivity in preference to the student's particular style, and thus may focus on received knowledge at the expense of the process of constructing new understandings (p. 350). Presumably, the hegemonic discourse secures governance and students need to be trained in order to ensure a legitimate form of power/knowledge. Furthermore, according to feminist critics (Harding, Keller, Longino), these hegemonic (masculine) power relations lead to dualisms such as rational-emotional, logical-intuitive, objective-subjective, and abstracted-holistic. With this background, Hildebrand (1998) concluded: "As a feminist, I argue that both sides of these dualistic concepts are present in science and should be portrayed within the available texts in schools. Science involves both the logical and the intuitive, both the objective and subjective ... Similarly, constructions of gender will only become liberatory when we are all free to move between both rational and emotional, both abstracted and embedded thinking modes" (pp. 348–349, Classified as Level III). At the end of her article, she acknowledges that there is more than one right way to write in science classrooms. This clearly shows the importance of recognizing an interaction between the subjective and the objective.

Feminist pedagogy has been generally interested in investigating questions such as: what is it about existing science cultures and methods of inquiry that excludes women? However, some feminists have gone beyond by challenging our traditional understanding of scientific knowledge by encouraging students to question the underlying masculine biases of our culture regarding objectivity, truth and the scientific enterprise. With this background, Mayberry (1998) has suggested:

> Feminist scholars have also critically examined traditional scientific inquiry on other grounds, including its tendency to privilege the masculine, reinforce existing power structures, and *promote so-called objectivity while obscuring interactional and interdependent relations among natural and social phenomena ....* Underlying these issues is an educational concern: How can feminist scientists and educators incorporate a critical examination of science into courses in the natural sciences? (p. 450, italics added). Classified as Level III.

Indeed the agenda suggested by Mayberry goes beyond that of some feminist scholars. It even overlaps the current interest in history and philosophy of science to provide students a critical appraisal of science and the underlying dynamics of scientific progress (cf. Matthews, 2014a, b; Niaz, 2009, 2016; Niaz & Rivas, 2016). This also shows that the underlying issues (history and philosophy of science) are the same for both female and male teachers.

Richmond, Howes, Kurth, and Hazelwood (1998) designed courses for undergraduate and graduate teacher education classes to enable these students to have a critical understanding of how science has been narrowly and powerfully shaped and thus marginalized significant groups including women. Teachers were exposed to the feminist literature (based on the writings of Brickhouse, Harding, Keller, and Longino) in order to facilitate a more honest perspective of science instead of placing it on an "epistemological pedestal." Based on this experience the authors concluded:

> Feminist critiques of science and its uses in society have helped us to understand why science has been inaccessible to many, and why we often feel disconnected from the learning and practice of it. Feminist philosophers, historians, and sociologists of science have demonstrated how science has grown out of a Western male tradition that celebrates objectivity, distance, power, and technological progress, and is often used to support social injustice and the status quo. These writers have illustrated not only the foolishness, but also the danger, of thinking that science is objective and value free. The very fact that these characteristics are impossibilities—particularly when we insist otherwise—make the study of the natural world infinitely more attractive (p. 916). Classified as Level III.

An important finding of this study is that claiming that science is based on "objectivity" and is value free has led to the marginalization of important parts of the society. Reversing this trend can incorporate those who have been left out and what is more important the study of science can become more attractive. Indeed, emphasizing objectivity with no reference to its evolving nature and that science is value-free can deprive many students to understand how science has progressed.

4.2.11 Indigenous Worldviews and Objectivity

In order to understand the contributions of the Yupiaq culture (southwestern Alaska) to science, Kawagley et al. (1998) first reproduce the following quote:

> The sciences accounted for in this book are largely part of a tradition of thought that happened to develop in Europe during the past 500 years—a tradition to which people from all cultures contribute today. (Rutherford & Ahlgren, 1990, p. 136)

Despite recognition by Rutherford and Ahlgren that today (as in the past) all cultures contribute to science, Kawagley et al. (1998) contend that Western science has become the prototype for what counts as science today and other ways of thinking and doing science have been largely ignored by the Euro-American scientific communities. This tendency to define science strictly from the viewpoint of Western culture has serious and detrimental ramifications for students from non-Western (including indigenous) cultures and languages:

> With its emphasis on controlled experimentation, replicability, and *alleged objectivity*, science as practiced in laboratories and as traditionally taught in U.S. schools does differ from the practice and thinking found in many indigenous cultures, but does that mean that what occurs in other cultures is not truly science? Our experience with the Yupiaq culture in southwestern Alaska leads us to believe that such indigenous groups practice science in ways that has similarities to—and important and useful differences from—Western science, and that the worldview underpinning this indigenous vision of science has valuable implications for science instruction (Kawagley et al., 1998, p. 133, italics added). Classified as Level III.

It is important to note that in contrast to Western science (largely conducted in laboratories), Yupiaq science is based on observation of the natural world coupled with direct experimentation in the natural setting. Knowledge from different sources (Yupiaq and Western) could perhaps be integrated in order to provide a deeper understanding of the world that surrounds us.

4.2.12 Nature of Science and Objectivity

Matthews (1998) traces the long history of writings that established the cultural, educational, and scientific benefits of teaching about the nature of science (NOS) based on a history and philosophy of science framework. However, during the past three decades, questions about the nature of science have become both more contentious and more pressing than they were previously:

> There had been a degree of cultural and philosophical unanimity about the nature and purpose of science. Of course, there was some dispute about the topic—inductivists versus falsificationists, positivists versus realists, Kuhnians versus Popperians, empiricists versus rationalists, Bernalian state-interventionists versus Polyanian free enterprises, etc.—but these disputes were basically domestic ones. There was general agreement that science was a good thing, that it was a cognitive enterprise abiding by intellectual standards, that it valued objectivity, that it sought to find truths about the world, and that it

gave us the best possible understanding of nature and reality. Merton's characterization of science as open-minded, universalist, disinterested, and communal (Merton, 1942) summed up professional and lay opinion on the matter (p. 162, italics added). Classified as Level I.

This account succinctly summarizes the debates in the history and philosophy of science community starting from about the middle of the twentieth century. Needless to say, science educators were also a part of these developments. During the 1980s the issues in science education became more contentious (as suggested by Matthews), primarily due to the introduction of radical constructivism by von Glasersfeld. Although, Merton's characterization of science is still important for science educators, philosophy of science itself has explored new territory. For example, Giere (2006a, p. 95) recounts how Newton's gravitational theory has been superseded by Einstein's theory of relativity, and consequently it is *presentist hubris* to think that we can have an objectively correct or true theories. With this background it is easy to understand how this presentation was classified as Level I, primarily because it emphasizes objectivity to be a part of the traditional scientific outlook.

A review of the history of science education shows that distinct goals of science education have been related to the larger goal of scientific literacy. DeBoer (2000) has argued that scientific literacy should be conceptualized broadly enough so that individual classroom teachers can pursue the goals that are most suitable for their particular situations along with the content and methodologies that are most appropriate for them and their students (p. 582). According to the author, science is a particular way of looking at the natural world and recommended that:

> Students should be introduced to this way of thinking and learn how to use it themselves since it is such an important means of generating knowledge of our world. Students should also be able to recognize when the methods of science are used correctly by others and when they are not. The validity of data, the nature of evidence, *objectivity and bias*, tentativeness and uncertainty, and assumptions of regularity and unity in the natural world are all important concepts for students to be aware of. At the same time, students need to recognize the limits of science and the power of other ways of thinking that are also functional in the world (DeBoer, 2000, p. 592, italics added). Classified as Level II.

Indeed, the recognition of *objectivity and bias* as an important topic (among others) for teaching scientific literacy opens the possibility for teachers to discuss it depending on the subject being discussed in class. Furthermore, a discussion in the classroom with respect to, whether the methods of science have been used correctly by scientists, can provide students an opportunity to understand "science in the making" (cf. Niaz, 2012).

The difference between the ecological understanding of science professionals (scientists and technicians) and non-scientists (adults and middle school students) has been investigated by Hogan and Maglienti (2001). Quantitative and qualitative analyses revealed that the participants' responses differed especially with respect to their emphasis on criteria: empirical consistency for science practitioners versus plausibility of the conclusions for the non-scientists. According to the authors this difference in epistemic criteria also helps to understand the difference between expert and novice scientific reasoning and concluded:

> Concluding that the students lacked expertise as scientific reasoners when they judged conclusions using epistemic criteria of coherence with their personal theories or inferences risks holding them to an outmoded, positivist standard of rationality that *emphasizes objectivity through the application of theory-independent rules of inference.* In contrast, postpositivist perspectives in the philosophy of science (e.g., Boyd et al., 1990) acknowledge the intricate interplay of theory and methodology in science. Cognitive scientists portray this interplay as a problem solving process that involves the recursive search of two problem spaces: the experiment space of data and methods, and the hypothesis space of conjectures and theories (Klahr, Fay, & Dunbar, 1993). Domain-specific information about the plausibility of hypotheses influences scientists' search of both hypothesis and experiment spaces, whereas domain-general heuristics constrain and guide their explorations (Hogan & Maglienti, 2001, p. 681, italics added). Classified as Level III.

This statement highlights that non-scientists as novices can draw conclusions that cohere with their personal theories or prior knowledge and this recognition can even help them to evaluate their views critically while evaluating new information. Despite this recognition to attribute lack of scientific reasoning to the thinking of novices amounts to a positivist understanding of objectivity. Next these authors emphasize the intricate relationship between data and their interpretation through conjectures and theories. Based on this, cognitive scientists (Klahr et al., 1993) recognize the importance of both types of knowledge, namely domain-specific (plausibility of hypothesis) and domain-general (heuristics that guide investigations). The importance of domain-specific and domain-general aspects of nature of science (NOS) have been the subject of considerable controversy in the science education literature. For example, Duschl and Grandy (2013) have explicitly endorsed that science educators need to explore only domain-specific aspects of NOS. In contrast, Niaz (2016) has reasoned that for meaningful learning we need to integrate the domain-specific and domain-general aspects of NOS, which coincides with the recommendation of the cognitive scientists.

Chen et al. (2013) developed an empirically based questionnaire to monitor young students (sixth graders) conceptions of nature of science (NOS). The questionnaire entitled, "Students' Ideas about Nature of Science (SINOS)" measured views (among others) on: (1) theory ladenness; (2) creativity and imagination; (3) tentativeness of scientific knowledge; (4) durability of scientific knowledge; (5) coherence and objectivity in science. These authors consider the first three constructs to be subjective in which scientists give importance to non-rational factors in the development of science, such as religion, culture, metaphysical beliefs, creativity, and imagination. The last two constructs are considered to be related to scientific objectivity as they highlight a stable opinion formed regarding a topic and consistency between experimental results and theories. According to these authors, scientific objectivity thus achieved can be attributed to Hacking (1983) and Ladyman (2002). The reference to Hacking (1983) in this context is interesting as the next chapter will discuss the relationship between representation and intervention, as suggested by Hacking. Furthermore, constructs in the subjectivity category correlated with science achievement (correlation coefficients in the range of 0.13–0.22), whereas constructs in the objectivity category correlated negatively. These findings imply that the objectivity category did not contribute to achievement scores as it represented a body of facts for memorizing. These are interesting findings and need to be researched further. At this stage, it is important to note how the authors summarized their results:

> Extremes of subjectivity or objectivity are not desirable. Science is not totally irrational, nor is it completely objective. During the development of a scientific idea or investigation of scientific research, *views in the subjectivity category interact with and are counterbalanced by views in the objectivity category, and vice versa*, by means of communications in the science community such as peer reviews and publications. Views in both categories play important roles in the construction of scientific knowledge, and may therefore influence science learning. (Chen et. al., 2013, p. 415, italics added)

This presentation was classified as Level IV as it suggests that the interaction between subjectivity and objectivity forms an integral part of scientific development, and is characterized by a dynamic relationship between the two "extremes." In a sense this approximates to what Daston and Galison (2007) consider as the reason why scientists started to abandon mechanical objectivity and embraced trained judgment, and that the latter did not replace the former, but on the contrary the two complement each other.

In traditional laboratory work students are, to a large extent, occupied by doing measurements and manipulating apparatus. This does not provide them with an opportunity to verbalize theoretical knowledge and relate theory to practice. Havdala and Ashkenazi (2007) designed a laboratory study in which they took the emphasis away from Israeli freshman chemistry students' hands-on work and, instead focused on pre-and post-lab activities. Students were interviewed with respect to their views about science and the following is an example of an empiricist-oriented view:

Q: How would you define a scientific law?

A: It is something experimental, for sure. It is not God-given, but it is something you can trust to be correct in 99%.

Q: What is it based upon?

A: Only on experiment (Reproduced in Havdala and Ashkenazi, 2007, p. 1141).

Based on the data from the interviews, students were classified into the following groups: empiricist-oriented, rationalist-oriented, and constructivist-oriented. Students' views about science were correlated with their approaches to lab practice. A coherent epistemological theory was constructed for each case, by considering the different degrees of certainty and confidence each student attributed to theoretical versus experimental knowledge in science. Finally, the authors concluded:

> The three epistemological theories, which correspond to empiricist-, rationalist-, and constructivist-oriented views, cannot form a single continuum of development from naïve to informed views about NOS. It is highly unlikely that an empiricist-oriented student, who holds objective views about empirical evidence and views theory as subjective, will switch to a rationalist-oriented view, which regards inference as objective and observation as subjective, or vice versa. Therefore, there are at least two different paths on which a student can progress from a completely naïve view, which equally trusts all knowledge, to

an informed view, which sees the subjective limitations in both components of knowledge and evaluates scientific knowledge by coordinating theory and empirical evidence (Havdala & Ashkenazi, 2007, p. 1156). Classified as Level III.

Interestingly, such encounters in the classroom can provide an opportunity to understand the objectivity–subjectivity continuum.

4.2.13 Postmodernism and Objectivity

As Editor of the *Journal of Research in Science Teaching*, Ron Good expressed concern with respect to the following statement by the National Research Council's, *National Science Education Standards: a sampler*: "The National Science Education Standards are based on the postmodern view of nature of science" (NRC, 1992, p. A-2). This may be cause of concern also for many science educators, and Good (1993) provided the following advice: "To question the *objectivity* of observation or the *truth* of scientific knowledge, one does not need to travel to the wispy world of postmodernism. Logical positivism and postmodernism are at the extremes of a long continuum of positions taken by scholars of the nature of science. It is not necessary to carry along the unwanted (unwarranted) baggage of either logical positivism or postmodernism to place oneself, as did the authors of *Science For All Americans*, in a more 'scientifically' defensible position" (p. 427). Among postmodern philosophers, Good specifically mentions Paul Feyerabend and Michael Foucault. However, it is important to note that this was written many years ago and recent scholarship in the field (Daston & Galison, 2007) has facilitated a better understanding of the development of objectivity in the history of science (especially see a discussion of Feyerabend's views in Chap. 6).

Constructivism and relativism in science pose considerable difficulties for science education. Some science educators have endorsed a relativist image of science that approximates to that of postmodernism (Bencze & Hodson, 1999; Roth, 1995). Most scientists have ignored the postmodern debate about science. However, some have deplored that the challenge to the legitimacy of science has been appearing within the science education movement itself (Holton, 1996, p. 552). Harding and Hare (2000) have argued that it is preferable to be open-minded rather than embrace relativism:

Individual scientists make errors just like everybody else and are not always objective, but with many individuals, each making some errors but each repeating and rechecking the work of others, the errors are corrected and the final result is much more accurate. This makes the community of scientists more effective than individuals .... The same fact can be reconsidered and rejected any time later. Indeed, if scientists thought decisions were final, they would probably be less willing to make them in the first place and closure would probably disappear. Whereas scientists have good reason to accept certain theories as true, they always do so with the proviso that they can change their minds later if new evidence warrants; this fits the definition of open-mindedness, not relativism (p. 229). Classified as Level IV.

Some observers and especially school science give the impression that closure in science marks a boundary and a fact is never again questioned. This clearly shows the role played by the scientific community in correcting knowledge claims and that the scientific attitude is characterized by open-mindedness and even if a theory is considered as correct today it can be changed later.

4.2.14 Science as a Career for Women and Objectivity

Baker and Leary (1995) have reported that women liked learning science in an interactive social context rather than participating in activities that isolated them such as independent reading, writing, or note taking. Many participants in their study were drawn to science careers due to some strong affective experiences that emphasize relationships and connectedness to the objects of study and the members of their research teams. Women practice their craft that emphasizes:

> This feeling of connectedness to nature has led to breakthroughs in fields such as primatology and genetics, but as many of the women scientists interviewed by Sheperd (1993) reported that it also leads to conflict with mentors and colleagues, isolation, and slower rates of promotion. In extreme cases, the conflict between doing science in a related and connected way and the norms of science that emphasize hierarchy, distance, and objectivity, lead to dropping out of science (Baker & Leary, 1995, p. 5). Classified as Level III

This clearly shows how objectivity although considered as an epistemic virtue can even constitute a stumbling block for some future scientists.

Cronin and Roger (1999) developed a conceptual framework of career opportunities for women in science, engineering and technology (SET), in Scotland and reported that:

> During the past 20 years, a rising tide of critical analysis has challenged science's claims to objectivity and neutrality. There exist many criticisms of the contention that the scientific community is representative of the gender, racial, or class diversity in society, and therefore of the possibility that it can be objective and neutral, or that the context in which studies are undertaken is neutral .... Drawing on work by Harding (1987) and Hubbard (1988), Duran stated it bluntly: "contemporary science's failure to acknowledge that it, too, is driven by social forces beyond its control and is responsive to social conditions that it pretends to ignore leaves us with science-as-lie" (1991, p. 92). Classified as Level III.

This study deals with a twofold problem: first, the scientific community generally does not recognize that science is socially embedded and second, to claim that science is objective and neutral is also questionable when we study the historical evolution of objectivity in the history of science. With this background, Cronin and Roger (1999) concluded that the underrepresentation of women in SET continues to be both progressive and persistent (p. 637).

4.2.15 Science in the Making and Objectivity

According to Harding and Vining (1997), it is not important to teach students the methods of science (science in the making) but rather a framework of knowledge.

As an example they provide the following changes in scientific thinking: Theory of evolution (1858) → Genetics (1900) → Molecular biology (1953). From the perspective of Harding and Vining, it is not important to teach students as to why these changes in theories took place, and textbooks can easily explain just the theories and ignore the distracting details. Boulton and Panizzon (1998) have criticized this approach on the grounds that, "… it is important that students understand *how* the knowledge that they learn is earned, why data may remain unchanging but scientists' conclusions might vary, and most important, that there is considerable subjectivity in science—after all, researchers are human" (p. 475, original italics). Furthermore, they emphasize that the scientific process must be taught progressively, so that by the time students complete their secondary school, "… they can appreciate why scientific information varies from text to text, why most science is collaborative, and why subjective decisions and serendipity may play major roles in a field reputed to be objective" (p. 476, Classified as Level III). At this stage it is important to clarify that (Boulton & Panizzon, 1998) suggesting that we present to the students the methods of science or the processes of scientific investigation does not imply that we teach "a glib rendition of the traditional scientific method" (p. 477). Actually, it is the *methods* by which the data are obtained that dictate the results and hence the conclusions. In the determination of the elementary electrical charge (oil drop experiment), the controversy between Robert Millikan and Felix Ehrenhaft was primarily based on the methods they used to handle data and consequently the different interpretations (for details see Holton, 1978a, b; Niaz, 2005). Interestingly, most general chemistry and physics textbooks not only ignore the method, but the controversy itself and thus perpetuate a myth regarding the method used by Millikan (cf. Niaz, 2015).

Yore, Hand, and Florence (2004) have emphasized that an understanding of scientists' ontological and epistemic beliefs about science is essential to understand their scientific practice, especially with respect to doing, writing, and reviewing research. Among others, these authors formulated the following research questions: (a) What views of the nature of science (NOS) are held and used by academic researchers?; and (b) Are academic researchers' beliefs about the NOS reflected in their beliefs about writing strategies and processes? The study is based on 19 university faculty members, working in science-related disciplines (Biology, Biochemistry, Microbiology, Chemistry, Computer science, Earth and Ocean sciences, and Astronomy) at a midsize Canadian university. All participants were interviewed and responded to a science writer questionnaire. Based on the data obtained, it was found that all of the scientists rejected the extremes of the traditional absolutist and postmodern relativist view. Instead the scientists positioned themselves around the modern evaluativist view of science according to which scientific knowledge is a temporary explanation that best fits the existing evidence and current thinking. Based on their findings the authors concluded:

Some scientists, for example, may choose to use the passive voice to stress their *objectivity*, knowing full well that they were personally involved in the actions. A few respondents expressed concerns about the discrepancy between the evaluativist view of science and the traditions and conventions of some academic journals. They believed that: a) active voice more clearly illustrated the personal involvement of scientists in the construction of the knowledge claims; and b) the lived experiences and the personal perspectives of the writers needed to be recognized by the audience and used by them to evaluate the creditability of claims in text. Both scientists and engineers ascribed priority to audience in making decisions about writing and text (Yore, Hand, & Florence, 2004, p. 365, italics added). Classified as Level III.

This clearly shows that some participants perceived a conflict with respect to reporting research in an active or passive voice. One scientist expressed the view that writing in the passive voice can be used dishonestly. Another expressed the view that the use of entirely passive voice was "outmoded" and favored the mixing of passive voice (Methods section) and active voice (Discussion section). The use of active voice potentially recognizes the human dimension in data interpretation and knowledge construction. However, the influence exerted by the editors and the audiences of the academic journals has been recognized in scientific writing (cf. Medawar, 1967).

4.2.16   Scientific Arguments and Objectivity

In the literature of the history and philosophy of science, scientific arguments and their construction have been interpreted through two different theoretical perspectives. The internalist perspective assumes that arguments can be evaluated on the basis of their internal consistency and their deductive and logical method of accounting for the majority of observed events. Furthermore, this perspective presupposes that scientists are objective observers and seek to produce knowledge that represents the world as it "truly" is. On the other hand, the externalist perspective considers that science is a very human activity and is consequently influenced by competition, bias, rivalry, and other fallible human characteristics that emphasize the importance of the tentative, creative, subjective, and evolving nature of the arguments constructed within the scientific community. Based on this perspective, Yerrick (2000) concluded:

> Claims of scientific *objectivity*, truth, and the production of facts and theories have undergone close scrutiny. Sociolinguistic and ethnographic studies of scientific settings … have provided insights into objective claims about the world …. For example, scientific laboratory methods of gathering and treating data may be changed due to some new finding, but the data and methods reported still appear polished within acceptable error, written in retrospect as highly rational and logical from their inception. Latour and Woolgar argued that a significant part of the work of scientists involves the transformation of observations and claims into scientific facts. They argued that scientific work is achieved largely by building arguments to persuade or convince other scientists through competition—a process that forces documents and data to fit particular outcomes for reasons other than pure rationality (p. 812, italics added). Classified as Level III.

This presentation coincides quite closely with what recent history and philosophy of science has researched and found with respect to scientific arguments and how these are constructed in the world of the real scientists subject to the difficulties involved in charting new territories and the ever-present peer pressure (Holton, 1978a, b). Both the internalist and the externalist perspectives can be discussed in class. However, most teachers would face a dilemma if the students go from one extreme (internalist) to another (externalist). One alternative would be to avoid the discussion of such controversial issues. Yerrick (2000) suggests that neither perspective is necessarily accurate (p. 812). Daston and Galison (2007), on the other hand, would suggest that the very concept of "objectivity" has been evolving in the history of science. Furthermore, to avoid the discussion of controversial issues in the classroom can even lead to teaching a "sanitized" version of science. Construction of a scientific argument based on the sanitized version entails covering up the confusion, random, and chaotic means that produced it so as to give the impression that it is an objective reflection of the world as it really exists.

4.2.17   Scientific Method and Objectivity

In a study based on a reflective, explicit, activity-based approach, Akerson, Abd-El-Khalick, and Lederman (2000) facilitated pre-service elementary science teachers' (at a state university in Western USA), understanding of the following nature of science (NOS) aspects: empirical, tentative, subjective (theory-laden), imaginative and creative, social, and cultural. Based on an open-ended questionnaire, before the intervention, researchers found that:

(a) When participating teachers recognized that theory generation might involve creativity, they also noted that scientists still have to follow the scientific method to ensure their objectivity and following is an example of a response provided by one participant: "A good scientist must be creative to design a good experiment .... The scientist might be imaginative in coming up with a theory, but it must be through the scientific method so that they stay objective" (Reproduced in Akerson, Abd-El-Khalick, & Lederman, 2000, p. 308).

(b) When participants admitted a role for imagination and creativity in science, they restricted this role to the design stages of investigations. Participants noted that it was not acceptable or desirable to use creativity or imagination when interpreting data, as this could compromise the objectivity of the scientists. Following is one example of such a response: "A scientist only uses imagination in collecting data .... But there is no creativity after data collection because the scientist has to be objective" (Reproduced in Akerson, Abd-El-Khalick, & Lederman, 2000, p. 308).

These responses highlight the importance of the scientific method in the generation of a theory and especially those phases of the scientific endeavor that go beyond the collection of data. According to the perception of these pre-service elementary science teachers in some phases of the scientific enterprise (interpretation of data) lack of a scientific method may deprive the scientists of their objectivity. It would be interesting to explore this idea in science textbooks, namely the scientific method (a set of procedures outlined in the form of a flow-diagram) facilitates the scientists to be objective. In the next chapter, this idea will be the subject of criteria for evaluating general chemistry textbooks. These authors recognized that in order to facilitate an understanding of the complex relationship between the degree of subjectivity and how scientists grapple with such issues in order to approximate objectivity requires in depth study of some historical episodes: "The subtleties of the influences of subjective, and social and cultural factors on the generation of scientific knowledge or the work of scientists are hard to convey or capture in the absence of rich and extensive contextualization, such as historical case studies of the development of some scientific construct or discipline" (p. 312). Classified as Level III.

Science teachers in many parts of the world have alternative conceptions (misconceptions) with respect to various aspects of the nature of science (NOS) and one study expressed this in cogent terms:

> Many [science teachers] also do not recognize the role of subjectivity (theory-ladenness) and social and cultural influences on scientific knowledge development. Most believe scientists are particularly objective, and that *use of the scientific method in developing scientific knowledge ensures objectivity.* They do not appreciate the roles that background knowledge and cultural influences play on scientists' designs and interpretations of data. For example, they do not recognize that scientists with differing content knowledge levels or cultural backgrounds may have different interpretations of the data (Akerson, Morrison, & McDuffie, 2006, pp. 195-196, italics added). Classified as Level III.

It is plausible to suggest that various NOS aspects need to be introduced in the classroom in the context of domain-specific historical episodes that form part of the science curriculum (for details see Niaz, 2016).

4.2.18   Social Dimensions of Science and Objectivity

Publication of Desmond and Moore's (1991) biography of Charles Darwin has shown that his theory was inextricably linked with its social dimensions. Despite this connection some scholars consider Social Darwinism to be something extraneous that was added later to the Darwinian corpus. According to Duveen and Solomon (1994):

> No one would argue that evolution theory did not produce grave social effects, even if only indirectly. However, there have been suggestions that Darwin himself maintained the kind of *scientific objectivity and aloofness from the social consequences of his theory* that some educators still insist that all "good scientists" should. Social implications, such apologists argue, were drawn by others who were not scientists, and so it follows that these effects should play no part in our science teaching .... General historical and philosophical arguments (e.g., Collins, 1982; Fuller, 1988; Holton, 1978b) show that both the epistemology of science and the analogies available for developing theory are strongly dependent on social and cultural influences. *Objectivity in its purest sense is never an option* (p. 575, italics added). Classified as Level III.

Based on this background, these authors recommend that social dimensions of Darwinian theory be included in school science as this helps to understand nature of science by presenting the human aspects of the scientists' struggle to forge new theories. Interestingly, in order to arrive at this conclusion, these authors had recourse to a wide range of expert opinion, such as science studies (Harry Collins), social epistemology (Steve Fuller), and history of science (Gerald Holton). In a sense this endorses Daston and Galison's (2007) thesis of trained judgment. Furthermore, it is important to note that these authors seem to suggest that although some of form of objectivity is acceptable in scientific research, however, in its purest form, as suggested by the scientific method, it is not an option.

Abd-El-Khalick, Waters, and Le (2008) have elaborated criteria for evaluation of nature of science in high school chemistry textbooks (published in USA), and explicitly differentiated between two aspects, namely "social dimensions of science" and "social and cultural embeddedness of science":

> The first aspect specifically refers to conceptualizations of "science as social knowledge," which should not be confused with relativistic notions of scientific knowledge. It refers to conceptions of science as advanced by philosophers of science such as Helen Longino ... [and] serve *to enhance the objectivity of collectively scrutinized scientific knowledge through decreasing the impact of individual scientists' idiosyncrasies and subjectivities.* In comparison, the "social and cultural embeddedness of scientific knowledge" aspect refers to the impact of the interactions between science and the social and cultural milieu in which it is embedded on, for instance, the sort of research that is pursued ... (e.g., research related or perceived to be related to human cloning). (p. 839, italics added). (Classified as Level IV).

This presentation (see the part in italics) clearly conceptualizes how the work of a scientist may be affected by the personal idiosyncrasies and subjectivity. It is precisely the social interactions among members of the scientific community that leads to an evolving nature of objectivity. In other words, it is the social dimension of science (e.g., peer-review process) that facilitates the transition from a subjective to a more objective nature of scientific knowledge. Furthermore, the difference between "social dimensions of the scientific enterprise" and "social and cultural embeddedness of science" is important. Nevertheless, it is important to point out that once the conflicts with respect to the funding of a research project (e.g., AIDS, Ebola, human cloning) have been resolved the subsequent phases of the research once again are dependent on the idiosyncrasies and subjectivity of the scientists.

Scientific inquiry, according to Ebenezer, Kaya, and Ebenezer (2011), is characterized by the following hallmarks: (a) Scientific conceptualization involves the identification of and development of deeper understanding of core science concepts that are necessary to shape inquiry; and (b) Scientific investigation involves skills such as framing a relevant research question, evaluating design and using mathematical knowledge and representations. Based on these considerations authors suggest that, "Scientific communication involves the sharing of ideas with respect to research questions, methods, and claims for peer response and evaluation meeting objectivity from a social perspective" (p. 99). Situating objectivity in the context of peer response and a social perspective shows its problematic nature. Classified as Level III.

Fusco (2001) developed a community-based action research science project to understand what it means to create a practicing culture of science learning. The following question guided the research project: how can an urban planning and community gardening project help to create a learning environment in which science was relevant? Based on this experience the author concluded:

> The action research methodology challenges the Western science tradition of dualistically separating objective and subjective, researcher and researched, rational and emotive, knower and known .... Knowledge and the ways in which knowledge is produced do not emerge objectively but occur within specific cultural, historical, and sociopolitical contexts. Engaging participants in the production of knowledge toward socially responsible ends is the explicit objective of action research (p. 864, Classified as Level III).

In a sense such action research-based projects approximate to what a scientist does to understand a phenomenon or solve a problem that needs the exploration of new ideas. Engaging participants in the process of knowledge construction helps to break the mold that is such an important part of traditional classroom practice. It seems that learning and doing science require that participants go beyond being objective and rational.

Bianchini and Solomon (2003) explored the discursive and social practices of beginning science teachers in a course on the nature of science and issues of equity and diversity. The discussion was organized along three dimensions: personal, social, and political. It was found beginning teachers routinely drew from only one of these three dimensions (instead of drawing on all three) to support their views of nature of science. During the discussions, one of the participating teachers (Travis) questioned Harding's (1998) argument for the acceptance of Southern, Eastern, and indigenous knowledge systems as science on the grounds that she was presenting the Western view as a monolithic entity in complete alignment with the scientific world view and concluded: "So, to me it's not so clear that Westerners all hold this *objective Western way of science* that we're sort of defining. A lot of people believe in ghosts, believe in God, believe in all this stuff. How does that fit in here" (p. 68, italics added). The point Travis is trying to make is that Harding wants to elevate some aspects of non-Western ways of understanding to the level of science and still denies the same to some Western aspects, such as astrology and religion. This clearly shows the problematic nature of objectivity in understanding Western and non-Western ways of understanding science. Classified as Level III.

Kittleson and Southerland (2004) studied a group of students in a mechanical engineering senior capstone design course to document the interaction patterns and knowledge construction activities. A total of 20 lab sessions were audio taped, in which at least two students were working on a simulation or experiment. Data were analyzed using Gee's (1999) method of discourse analysis. In one of the sessions a group investigated automobile defrosters, which have been used for many years, yet little is known about the heat transfer and fluid dynamics phenomena related to them. Most of the time in lab (both simulation and experimentation) was spent on collecting various types of data. Members of the group believed their data reflected what was actually happening in the real world. In an interview, one of the students reported, "I mean, obviously the experiments are right because, well, they're the real stuff" (p. 284). For anyone familiar with history of science, data in themselves do not constitute scientific knowledge, but rather it is the interpretation of the data with support from the scientific community that facilitates "science in the making." According to the authors, such views are based on the assumption that one can determine what is really happening, and are congruent with objectivist epistemology and realist ontology. With this background the authors have discussed the role played by personal bias in scientific research and the universal character of science:

> Essentially, universalism of science suggests that scientific methods are powerful enough so that any inquirer in any society or culture would get the same results as another inquirer doing the same inquiry .... In this sense, science at the community level would be free from subjective influences. However, theorists such as Harding would critique *the degree to which objectivity could be achieved at the community level*. Perspectives from the philosophy of science, ones which make explicit the role of the social context in the construction of knowledge (e.g., Kuhn, 1970; Longino, 1990) render it problematic to consider scientific knowledge without also considering how the social context can become incorporated into that knowledge (p. 269, italics added). Classified as Level III.

This shows yet another facet of the problematic nature of objectivity. In other words, all scientific knowledge gets sanctioned by the scientific community through the peer-review system. However, in the long run this knowledge is subject to a continual critical appraisal and thus subject to further changes based on the social context or other relevant factors.

Social dimensions of science and the difficulty of achieving complete objectivity in science has been the subject of considerable research. Similarly, the discourses on race and gender raise the question of objectivity in compelling ways. The Swedish sociologist Gunnar Myrdal (1944/1962) showed convincingly how American racism and the biological determinism of the craniometrists and psychometrists were based on questionable science. Gould (1981, 1995), considered as an enfant terrible by the scientific establishment, tackled the problem of racism and sexism and was alarmed to see how quantitative data were marshalled to support certain preconceived ideas such as those of IQ racist Cyril Burt. In a similar vein, the feminist scholar Donna Haraway (1991) questioned how the conception of objectivity was enshrined in the scientific establishment, which she dubbed as the "god's eye view" or the "the view from nowhere." When the Enlightenment introduced democratic and egalitarian notions that militated against a hierarchical ordering of people, science proved particularly effective in overcoming this resistance, based on political, moral, or religious grounds. Based on these considerations, Norman (1998) concluded:

> *With its convincing claim to complete objectivity, institutional science succeeded in positioning itself beyond the reach of moral, political, or religious scrutiny.* The prestige of science was effectively pressed in the service of overcoming those tendencies within the wider society that opposed strategies of dominance and exclusion. The claims of the scientists were deemed timeless, beyond the contingencies of culture and history. The relative inaccessibility of science to would-be critics allowed science to legitimize race and gender inequality by providing an authoritative basis for their semantic encodement .... Science provided the objective evidence of the natural inferiority of women, homosexuals, the lower

classes, the colonized, and the enslaved. On the basis of this objective evidence, the Enlightenment dictum about the equality of all humans could be overridden (p. 367, italics added). Classified as Level III.

This account shows an interesting facet of the progress in science in the social arena, and how the concept of objectivity provided the necessary support to introduce the democratic ideals of the Enlightenment. However, what started as a fruitful relationship between science and the society, finally led science to be beyond criticism and its claims were considered to be "beyond the contingencies of culture and history" and in this objectivity played an important part. Furthermore, if we accept that the claims of science are timeless, it would be difficult to understand how and why an important feature of the nature of science is precisely the tentative nature of science (for details see Niaz, 2016).

For teachers who accept science as socially constructed, value-laden, and context-bound, it is important for the students to understand that human endeavor plays an important role in constructing the reality that surrounds us. In this context, queer theory promotes human emancipation by focusing on issues of power, justice, ideologies, gender, and race. Science textbooks are generally written to represent the particular set of paradigms to which the scientific community is committed and thus perpetuate normal science (Kuhn, 1970). Based on queer theory, Snyder and Broadway (2004) analyzed eight high school biology textbooks (published in the USA), with the following premise:

> Diversity in the nations' schools is both an exciting opportunity and a complicated challenge. How do teachers and textbooks make sure all voices are heard—and no one is silenced in the journey toward scientific literacy? The challenge for the empiricist, modernist science teacher is easy: Science is an epistemology that is objective, directly value neutral, and established independent of contentions of politics and culture …. Science is often envisioned as directly reflecting the truths in nature and therefore unquestionable (p. 619). Classified as Level III.

Based on their research, the authors concluded that science educators need to create a demand for textbooks that present science in an equitable, socially relevant context that reflects the diverse nature of science that meets the needs of all students. Most science textbooks instill an empiricist epistemology in which the task of the teacher is reduced to a simple relator of events with no effort to understand the underlying substructure that entails controversies, rivalries among scientists, and alternative interpretations of data. Similarly, research in science education shows that many science textbooks in most parts of the world are written with a similar empiricist perspective (cf. Niaz, 2014).

4.2.19 Socioscientific Issues and Objectivity

Reform efforts in science education have recognized that presenting science in the abstract is neither motivating nor inclusive of the majority of students (Ziman, 1994). Science-technology-society (STS) curricula that give science an accessible social context have developed in response. STS is a broad umbrella term that may include a wide range of ideas that address history, philosophy, sociology of science, and contemporary economic, social and political concerns. STS may be seen as part of the curriculum in its own right or as a supplement to the traditional science curriculum. Salter's Advanced Level Chemistry Course developed in the UK is one example of how to teach scientific knowledge, laws, and theories within a social context. However, there has been some resistance to such courses as teachers fear that extensive coverage of socioscientifc issues devalues the curriculum, alienates traditional science students and jeopardizes their own status as gatekeepers of scientific knowledge. Furthermore, socioscientific issues are devalued with respect to the masculinity of abstract science. Hughes (2000) has argued that gendering of science is socially constructed and not biologically determined, and concluded:

> The origins of the symbolic masculinity of science can be traced back to the 17th- and 18th-Century Enlightenment. As belief in the power of rationality began to supersede dogma and superstition, *hierarchical dualistic splits emerged that associated reason/emotion and objectivity/subjectivity with a male/female divide* (Fox-Keller, 1992, pp. 16–21). The persistence of gendered dichotomous thinking is evident in contemporary associations of physical science with masculine hard abstract rationality, and human and social sciences with a feminine, more subjective, or softer approach. The abstraction and objectivity of pure science have masculine connotations, whereas a contextual approach is associated with the feminine. Any consideration of STS is therefore readily caught up in these gender hierarchical binaries (p. 434, italics added). Classified as Level IV.

This clearly shows how inclusion of socioscientfic issues in the science classroom is controversial. The degree to which masculine connotations played a role in the development of science can vary and depend on how one perceives progress in science. Nevertheless, as suggested by Daston and Galison (2007), the evolving nature of objectivity has indeed played an important role in introducing mechanical objectivity in which rationality of the scientist was considered to be a major driving force in understanding science.

Sadler et al. (2006) have explored middle and high school science teachers' perspectives on the use of socioscientific issues (SSI) and on dealing with ethics in the context of science instruction in the USA. SSI are usually value-laden, and the juxtaposition of science and ethics can be uncomfortable not only for scientists but also teachers and students who define science in terms of objectivity. Based on these considerations these authors concluded: "Socioscientific issues frequently involve complex problems subject to scientific data as well as ethical considerations; therefore, efforts to preserve the oft-perceived objectivity of science by excluding values and ethics from the science classroom shelter students from the complexities of science as it is conducted in and applied to society" (p. 354, Classified as Level III). Some of the topics suggested by the teachers that involve ethics and values include: genetics, gene therapy, stem cell research, and garbage collection.

Writing, talking, and reading about science (especially socioscientific issues) have the potential to facilitate scientific literacy. However, writing scientific narratives are not traditionally associated with learning science, as science is generally portrayed as a source of objective knowledge. On the contrary, narratives are sub-jective accounts of human experience, a genre with which most students are not familiar. Tomas, Ritchie, and Tones (2011) designed a mixed methods study in which they investigated the learning experiences of ninth-grade Australian stu-dents as they participated in an online science-writing project on the socioscientific issue of biosecurity. Besides writing scientific narratives that integrated scientific information about biosecurity, students completed questionnaires and were inter-viewed based upon open-ended questions. This experience led the authors to conclude:

> Although the credibility of data generated from qualitative interviews may be questioned due to its lack of "objectivity" and inherent human interaction, this may be considered a strength, as interviews capture the subject's perspective of the phenomenon under study, and enable them to formulate their own conceptions of reality in a dialogue with the researcher .... In order to explore the factors students attributed to the improvements in their attitudes from the project, they were asked at interview a number of questions about their experiences and perceptions of the project (Tomas, Ritchie, & Tones, 2011, p. 889). Classified as Level III.

In view of the fact that this study found that students developed more positive attitudes toward science and science learning, raises the issue and necessity of including activities in the classroom that involve human interaction and student participation. Historical evolution of the concept of objectivity (Daston & Galison, 2007) shows that at various stages science itself encouraged interaction and com-munication between participating scientists and the scientific community (trained judgment is a particularly good example of such objectivity). Actually, school science lacks the vitality of investigation, discovery, creative invention, and narra-tive understanding. An effective way to bridge the gap between school science and what scientists actually do, that is, "*science in the making*" is through the inclusion of humanizing aspects of the history of science in the form of a story (Klassen, 2011).

4.2.20   *Teachers' Emotions and Objectivity*

One important aspect of teaching science that needs more attention is how teachers feel about their teaching. In this context, it is important to look more carefully at the emotions of science teaching, both negative and positive, and to use this knowledge to improve the working environment of science teachers. In most edu-cational systems, there are some implicit and explicit emotional rules based on the following: (a) A teacher should not express her/his emotions because emotions are biased and there is no place for them in teaching or learning science; (b) Science should be objective; (c) A teacher should teach science the way everybody else does in the school, namely teach to the test and teach children "scientific knowl-edge." In a 3-year ethnographic study, Zembylas (2002) has explored the positive and negative emotions of an experienced elementary science teacher, as she constructed her science pedagogy, curriculum planning, and relationships with children and colleagues. Data sources included participant observation, in-depth interviews (including family stories), field notes, diaries, and videotapes. Based on this experience the author concluded:

> Developing a conceptual framework that analyzes emotions in science education is challenging because in Western philosophy, science, and culture, emotion has been traditionally opposed to reason, truth, and the pursuit of *objective knowledge*. The assumption held, since the time of Socrates, that affect and emotion are irrational and cannot be studied scientifically (as all sciences should be) and Cartesian dualism of mind and body contributed to a false polarization of reason and emotion .... In addition, issues of affect and emotions have been usually associated with women and feminist philosophies, and they therefore have been excluded from the dominant rationalist structures as worthwhile knowledge. Such a notion of knowing is based on knowledge as a manifestation of rationality; thus, an experience that is usually emotional threatens the disembodied, detached, and neutral knower (pp. 81–82, italics added). Classified as Level III.

In this particular study the female elementary science teacher was deeply involved with her students in out-of-class science exploration projects. It also became apparent that these activities were not considered as part of standard science content knowledge as stipulated by the curriculum. The lack of emotional and social support from other colleagues created in her feelings of failure, anxiety, and powerlessness. The author clarifies that he claims no objectivity or authority for his interpretations of the events and experiences related by the science teacher. Nevertheless, he does claim that these interpretations provided grounds for further reinterpretations and conversations leading to more insight and richer understandings. Perhaps it is plausible to suggest that the scientific endeavor is in itself also an exercise of enriching experiences through interactions and feedbacks from the scientific community.

4.2.21 Teaching Evolution and Objectivity

According to Dobzhansky (1973): "Nothing in biology makes sense except in the light of evolution" (p. 125). Despite its central role in modern biology, teaching evolution remains a difficult and controversial subject in many countries. Most teachers would perhaps agree that the goal for students is to acquire knowledge about evolution. Cobern (1994) considers this answer as simplistic and understandable only within a *scientistic* view of science, which is a myth in school science:

> The myth is a scientistic view roughly embracing classical realism, philosophical materialism, strict objectivity, and hypothetico-deductive method. Though recognizing the tentative nature of all scientific knowledge, scientism imbues scientific knowledge with a Laplacian certainty denied all other disciplines, thus giving science an *a priori* status in the intellectual world. The certainty of scientism can make life easy for the science teacher. Scientism allows the teacher to say to students that this is the way things are, for science provides the *one* reliable source of objective knowledge (p. 585, original italics). Classified as Level III.

Cobern's concern lies with the fact that although students may seem to understand evolution they generally do not believe in it. In other words, we are faced with the dilemma: Science educators need to facilitate conceptual understanding and/or persuasion for belief. Furthermore, as learning takes place in a social context, controversial topics like evolution cannot ignore the significance of the cultural milieu. Interestingly, Cobern suggests that the issue of belief cannot be ignored, and that belief is the place where instruction should begin (p. 587). Finally, Cobern (1994) states, "Today's teacher of evolution faces a situation very much like Darwin presenting the *Origin of Species* to a public that historically held a very different view of origins" (pp. 587–588). This scenario is crucial in teaching not only evolution but also all controversial topics of the science curriculum. Leon Cooper (1992), Nobel Laureate in physics, has provided cogent advice to solve the dilemma: "A question often very puzzling to students is why such a thing was done at such a time. Frequently, the answer can only be given in the *milieu of the time*—the problems that seemed important, the opinions of the people involved" (p. xii, Preface, emphasis added). Cooper goes beyond by suggesting that if the Michelson-Morley experiment (late nineteenth century) had been done at the time of Copernicus (sixteenth century), its result would have no significance for the astronomers, as they considered the earth to stand still and at the center of the universe. It seems that reference to *milieu of the time* can help to facilitate a better understanding of the beliefs of students in a particular topic. At this stage it is important to note that students' beliefs are closely enmeshed with their alternative conceptions of a topic, which have been investigated intensively.

Smith (1994) has criticized Cobern's (1994) approach to teaching evolution that focuses on belief in evolution, on the grounds that students may understand the term *belief* as synonymous with faith, opinion, or conviction. Furthermore, it may lead the students to understand that accepting evolution is a matter of personal faith that has no evidential basis. Actually, the role of evidential basis based on empirical evidence is controversial even in the history of science. For example, J.J. Thomson and E. Rutherford had very similar experimental evidence (alpha particle experiments) and still there interpretations were entirely different and in part based on their prior beliefs, theories, models, or theoretical frameworks (for details on this and other historical episodes see Niaz, 2012, 2016). Prior theoretical beliefs play a crucial role in scientific progress and the controversy between Thomson and Rutherford lasted for many years although they were well known to each other and could easily have met over dinner and resolved the controversy. Based on his critique with respect to teaching biology, Smith (1994) concluded:

> Although the distinction between believing and accepting may be a subtle one for many, it is crucial to understanding the nature of science; moreover, drawing carefully the distinction between belief (or faith) in the absence of objective evidence and acceptance that is based on evidence provides an excellent opportunity for helping students to understand what science is. In my view, in fact, the primary reason for including evolution in the curriculum, other than the obvious value of a meaningful understanding as a basis for understanding the rest of biology, is that it provides the wonderful opportunity for addressing pervasive misconceptions about the nature of science (p. 595). Classified as Level III.

More recently, Laats and Siegel (2016) have argued that a student does not need to believe in evolution in order to understand its tenets and evidence. In other words, a student can be fully literate in modern scientific thought and still maintain contrary religious or cultural views. Both Laats (a historian) and Siegel (a philosopher of science) agree that as a science creationism is flawed. However, given that creationism represents a form of religious dissent it is important to disentangle belief from knowledge. Interestingly, in the history of science even scientists (fully literate in scientific thought) can ignore experimental evidence and continue to believe in their prior theoretical frameworks. One example (Thomson versus Rutherford) of such a case was cited above. Similarly, Robert Millikan provided experimental evidence to determine Planck's constant h based on Einstein's photoelectric equation and still rejected Einstein's theoretical framework and continued to believe in the classical wave theory of light (for details see Niaz, 2012). Indeed, the role played by empirical evidence in the history of science is much more complex and controversial. The historical evolution of objectivity itself as studied by Daston and Galison (2007) is a good representation of how empirical evidence was cast in different ways, depending on the epistemological orientation of the scientists involved.

Centrality of Darwinian theory to biological thought has been recognized in the literature (Gould, 1977; Mayr, 1982). However, science educators (in USA and other parts of the world) have faced considerable difficulties in teaching biological evolution to students from orthodox (Christian, Jewish & others) background. Jackson et al. (1995) have reported the difficulties faced by a science educator (from a secular-humanist background in the northern USA) in trying to communicate biological evolutionary theory to scientists, science educators, and science teachers in the religious-influenced culture in the southern USA. This experience shows the limitations of the cognitively oriented conceptual change theory. Instead the authors used a *heuristic inquiry approach* (Patton, 1990) in which an overtly personal and subjective viewpoint is acknowledged. Elaborating on the methodology used, Jackson et al. (1995) concluded:

> First, this topic [biological evolution] elicits strong emotional reactions in many people, including several of the researchers .... In such circumstances, the *use of a method which explicitly strives for objectivity is probably futile and definitely presumptuous*. Second, the primary researcher/first author was conscious of a profound initial ignorance of the religious points of view on the issues raised. This situation calls for a method which anticipates and values an adaptive process by which specific research questions and methods evolve in response to data gathered and analyzed earlier in the inquiry (p. 590, italics added). Classified as Level III.

This statement clearly shows the difficulties involved in doing research on topics in which both the participants and researchers have strong prior epistemological views that produce conflicting situations in the classroom. Indeed, the authors recognize that in such studies participants need to be considered as co-researchers as they posed incisive questions that provided a stimulus to reflect and reevaluate the basic assumptions and goals of the study. In a sense these findings can be seen as the two poles of the subjectivity–objectivity interface. In other words, based on his professional training in evolutionary biology the primary researcher thinks that he is being objective and at the same time in his interactions with the participants he is forced to understand their views and hence the need for a subjective understanding. At this stage it would be interesting to reproduce some excerpts from an interview with a scientist with a Ph.D. in evolutionary biology, who participated in the study (interviewer's comments or questions are not included):

> My earliest background was in phylogenetic analysis ... I actually *do* couch myself as a fundamentalist, although many people would deny that I am ... I do believe that the Bible is God's word, it is inerrant .... The Bible has been kept intact—there have been word changes, but God has kept the *meaning* intact .... My standing with God has nothing to do with my stand on evolution. There's still a tension, it doesn't resolve .... Don't get me wrong—I accept evolution ... I'm often asked how I can be a Christian and a scientist, or vice versa—scientists ask me, and my Christian friends ask me, and they all assume that I must be compromising both sides, but I'm not, really—I accept that evolution is the mechanism the God used to create life (Reproduced in Jackson et al., 1995, p. 599, original italics).

The conflict between being a Christian and a scientist, belief in God or evolution, evidently entails the two poles of the subjectivity versus objectivity dichotomy. Most biology teachers in different parts of the world face the same dilemma.

This chapter provides examples of research reported in the *Journal of Research in Science Teaching* that facilitate a wide range of perspectives with respect to objectivity. Conclusions based on these findings will be integrated with those from other chapters and presented in Chap. 7.

References

American Association for the Advancement of Science, AAAS. (1993). *Benchmarks for science literacy: project 2061*. Washington: Oxford University Press.

Bencze, L., & Hodson, D. (1999). Changing practice by changing practice: Toward a more authentic science and science curriculum development. *Journal of Research in Science Teaching, 36*, 521–539.

Beth, E. W., & Piaget, J. (1966). *Mathematical epistemology and psychology*. Dordrecht: Reidel.

Boyd, R. N., Gaspar, P. & Trout, J. D. (1990). *The philosophy of science*. Cambridge: MIT Press.

Campbell, D. T. (1988a). Can we be scientific in applied social science? In E. S. Overman (Ed.), *Methodology and epistemology for social science* (pp. 315–333). Chicago: University of Chicago Press. (first published in 1984).

Campbell, D. T. (1988b). The experimenting society. In E. S. Overman (Ed.), *Methodology and epistemology for social science* (pp. 290–314). Chicago: University of Chicago Press.

Charmaz, K. (2005). Grounded theory in the 21st century: applications for advancing social justice studies. In N. K. Denzin & Y. S. Lincoln (Eds.), *The Sage handbook of qualitative research* (3rd ed., pp. 507–535). Thousand Oaks, CA: Sage Publications.

Collins, H. M. (1982). Tacit knowledge and scientific networks. In B. Barnes & D. Edge (Ed.), *Science in context*. Buckingham: Open University Press.

Cooper, L. N. (1992). *Physics: structure and meaning*. Hanover: University Press of New England.

Crabtree, B. F., & Miller, W. L. (1999). *Doing qualitative research*. Thousand Oaks: Sage.

Daston, L., & Galison, P. L. (1992). The image of objectivity. *Representations, 40*(special issue: seeing science), 81–128.

Daston, L., & Galison, P. (2007). *Objectivity*. New York: Zone Books.

Denzin, N. K., & Lincoln, Y. S. (2005). Introduction: the discipline and practice of qualitative research. In N. K. Denzin & Y. S. Lincoln (Eds.), *The Sage handbook of qualitative research* (3rd ed., pp. 1–32). Thousand Oaks: Sage.

Desmond, A., & Moore, J. (1991). *Darwin*. London: Michael Joseph.

Dobzhansky, T. (1973). Nothing in biology makes sense except in the light of evolution. *The American Biology Teacher, 35*, 125–129.

Duschl, R. A., & Grandy, R. (2013). Two views about explicitly teaching nature of science. *Science & Education, 22*(9), 2109–2139.

Fox-Keller, E. (1992). *Secrets of life, secrets of death: essays on language, gender and science*. London: Routledge.

Fuller, S. (1988). *Social epistemology*. Bloomington, IN: Indiana State University Press.

Freire, P. (1971). *Pedagogy of the oppressed*. New York: Continuum Books.

Gee, J. (1999). *An introduction to discourse analysis*. New York: Routledge.

Giere, R. N. (2006a). Perspectival pluralism. In S. H. Kellert, H. E. Longino & C. K. Waters (Eds.), *Scientific pluralism* (pp. 26–41). Minneapolis: University of Minnesota Press.

Gipps, C. (1999). Socio-cultural aspects of assessment. In A. Iran-Nejad & P. D. Pearson (Eds.), *Review of research in education 24*, (355–392). Washington: American Educational Research Association.

Giroux, H. (1992). *Border crossings: cultural workers and the politics of education*. New York: Routledge.

Gooday, G., Lynch, J. M., Wilson, K. G., & Barsky, C. K. (2008). Does science education need the history of science? *Isis, 99*, 322–330.

Gould, S. J. (1977). *Ever since Darwin*. New York: Norton.

Gould, S. J. (1981). *The mismeasure of man*. New York: Norton.

Guba, E. G., & Lincoln, Y. S. (1989). Fourth generation evaluation. Newbury Park: Sage.

Habermas, J. (1972). Knowledge and human interests. (trans: Shapiro, J.J.). London: Heinemann.

Hacking, I. (1983). *Representing and intervening*. Cambridge: Cambridge University Press.

Haraway, D. J. (1991). *Simians, cyborgs, and women: the reinvention of nature*. New York: Routledge.

Harding, S. (1987). *The science question in feminism*. Ithaca: Cornell University Press.

Harding, S. (1998). *Is science multi-cultural? Postcolonialisms, feminisms, and epistemologies*. Indianapolis: Indiana University Press.

Harding, P. A., & Vining, L. C. (1997). The impact of the knowledge explosion on science education. *Journal of Research in Science Teaching, 34*, 969–975.

Hodson, D. (1993). In search of a rationale for multicultural science education. *Science Education, 77*, 685–711.

Holton, G. (1969). Einstein and the 'crucial' experiment. *American Journal of Physics, 37*, 968–982.

Holton, G. (1978a). Subelectrons, presuppositions, and the Millikan-Ehrenhaft dispute. *Historical Studies in the Physical Sciences, 9*, 161–224.

Holton, G. (1978b). *The scientific imagination: case studies*. Cambridge: Cambridge University Press.

Holton, G. (1996). Science education and the sense of self. In P. R. Gross, N. Levitt & M. W. Lewis (Eds.), *The flight from science and reason* (pp. 551–560). New York: New York Academy of Sciences.

Hubbard, R. (1988). Some thoughts about the masculinity of natural science. In M. M. Gergen (Ed.), *Feminist thought and the structure of knowledge* (pp. 1–15). New York: New York University Press.

Keller, E. F. (1985). *Reflections on gender and science*. New Haven: Yale University Press.

Kitchener, R.F. (1986). *Piaget's theory of knowledge: Genetic epistemology and scientific reason.* New Haven, CT: Yale University Press.

Klahr, D., Fay, A. L., & Dunbar, K. (1993). Heuristics for scientific experimentation: a developmental study. *Cognitive Psychology, 25,* 111–146.

Klassen, S. (2011). The photoelectric effect: reconstructing the story for the physics student. *Science & Education, 20*(7–8), 719–731.

Kuhn, T. (1962). *The structure of scientific revolutions.* Chicago: University of Chicago Press.

Kuhn, T. (1970). *The structure of scientific revolutions* (2nd ed.). Chicago: University of Chicago Press.

Laats, A., & Siegel, H. (2016). *Teaching evolution in a creation nation.* Chicago: University of Chicago Press.

Lacey, H. (2004). Is there a significant distinction between cognitive and social values? In P. Machamer & G. Wolters (Eds.), *Science, values and objectivity* (pp. 24–51). Pittsburgh: University of Pittsburgh Press.

Ladyman, J. (2002). *Understanding philosophy of science.* New York: Routledge.

Lincoln, Y.S., & Guba, E.G. (2000). Paradigmatic controversies, contradidtions, emerging confluences. In N.K. Denzin & Y.S. Lincoln (Eds.), *Handbook of qualitative research* 2nd ed. (pp. 163–188). Thousand Oaks, CA: Sage.

Longino, H. E. (1990). *Science as social knowledge: values and objectivity in scientific inquiry.* Princeton: Princeton University Press.

Machamer, P., & Wolters, G. (2004). Introduction: science, values and objectivity. In P. Machamer & G. Wolters (Eds.), *Science, values and objectivity* (pp. 1–13). Pittsburgh: University of Pittsburgh Press.

Mayr, E. (1982). *The growth of biological thought: Diversity, evolution and inheritance.* Cambridge, MA: Belknap Press of Harvard University Press.

Medawar, P. B. (1967). *The art of the soluble.* London: Methuen.

Merton, R.K. (1942). Science and technology in a democratic order. *Journal of Legal and Political Sociology, 1.* Reprinted as 'Science and Democratic Social Structure', in his *Social theory and social structure.* New York: Free Press (1957).

Myrdal, G. (1944/1962). *An American dilemma: the negro problem and modern democracy.* New York: McGraw-Hill.

National Research Council, NRC (1992). *National science education standards: A sampler.* Washington, DC: National Academy Press.

National Research Council, NRC. (1996). *National science education standards.* Washington: National Academy Press.

Niaz, M. (1991). Role of the epistemic subject in Piaget's genetic epistemology and its importance for science education. *Journal of Research in Science Teaching, 28,* 569–580.

Niaz, M. (1997). Can we integrate qualitative and quantitative research in science education? *Science & Education, 6,* 291–300.

Niaz, M. (1998). From cathode rays to alpha particles to quantum of action: a rational reconstruction of structure of the atom and its implications for chemistry textbooks. *Science Education, 82,* 527–552.

Niaz, M. (2009). *Critical appraisal of physical science as a human enterprise: dynamics of scientific progress.* Dordrecht: Springer.

Niaz, M. (2011). *Innovating science teacher education: a history and philosophy of science perspective.* New York: Routledge.

Niaz, M. (2012). *From 'Science in the Making' to understanding the nature of science: an overview for science educators.* New York: Routledge.

Niaz, M. (2014). Science textbooks: the role of history and philosophy of science. In M. R. Matthews (Ed.), *International handbook of research in history, philosophy and science teaching* (pp. 1411–1441). Dordrecht: Springer.

Niaz, M. (2015). That the Millikan oil-drop experiment was simple and straightforward. In R. L. Numbers & K. Kampourakis (Eds.), *Newton's apple and other myths about science* (pp. 157–163). Cambridge: Harvard University Press.

Niaz, M. (2016). *Chemistry education and contributions from history and philosophy of science.* Dordrecht: Springer.

Niaz, M., Abd-El-Khalick, F., Benarroch, A., Cardellini, L., Laburú, C. E., Marín, N., Montes, L. A., Nola, R., Orlik, Y., Scharmann, L. C., Tsai, C.-C., & Tsaparlis, G. (2003). Constructivism: defense or a continual critical appraisal — a response to Gil-Pérez, et al. *Science & Education, 12,* 787–797.

Niaz, M., Aguilera, D., Maza, A., & Liendo, G. (2002). Arguments, contradictions, resistances and conceptual change in students' understanding of atomic structure. *Science Education, 86,* 505–525.

Niaz, M., & Robinson, W. R. (1993). Teaching algorithmic problem solving or conceptual understanding: role of developmental level, mental capacity, and cognitive style. *Journal of Science Education and Technology, 2,* 407–416.

Nurrenbern, S. C., & Pickering, M. (1987). Concept learning versus problem solving: is there a difference? *Journal of Chemical Education, 64,* 508–510.

Ogbu, J. (1978). *Minority education and caste: the American system in cross-cultural perspective.* New York: Academic Press.

Pascual-Leone, J., Goodman, D., Ammon, P., & Subelman, I. (1978). Piagetian theory and neo-Piagetian analysis as psychological guides in education. In J. M. Gallagher & J. A. Easley (Eds.), *Knowledge and development 2,* (243–289). New York: Plenum.

Patton, M. Q. (1990). Qualitative evaluation and research methods. Newbury Park: Sage.

Piaget, J. (1971). *Biology and knowledge: an essay on the relations between organic regulations and cognitive processes.* Chicago: University of Chicago Press.

Piaget, J. (1977). *Equilibration of cognitive structures.* New York: Viking.

Roth, W.-M. (1995). *Authentic school science.* Dordrecht: Kluwer Academic.

Rutherford, E. (1911). The scattering of alpha and beta particles by matter and the structure of the atom. *Philosophical Magazine, 21,* 669–688.

Rutherford, F. J., & Ahlgren, A. (1990). *Science for all Americans.* New York: Oxford University Press.

Sewell, Jr., W. H. (1992). A theory of structure: duality, agency, and transformation. *American Journal of Sociology, 98*(1), 1–29.

Sheperd, L. (1993). *Lifting the veil: the feminine side of science* Boston: Shambhala Publications.

Tashakkori, A. & Teddlie, C. (2003). *Handbook of mixed methods in social and behavioral research.* Thousand Oaks: Sage.

Tsaparlis, G. (2014). Linking the macro with the micro levels of chemistry: demonstrations and experiments that can contribute to active/meaningful/conceptual learning. In I. Devetek & S. A. Glažar (Eds.), *Learning with understanding in the chemistry classroom* (pp. 41–61). Dordrecht: Springer.

Wertheim, M. (1995). *Pythagoras' trousers.* New York: W.W. Norton.

Yeany, R. H. (1991). Dissemination and implementation of research findings: impacting practice. *NARST News, 33*(4), 1.

Ziman, J. (1994). The rationale of STS education is in the approach in science education. In J. Solomon & G. Aikenhead (Eds.), *STS education: international perspectives on reform* (pp. 21–31). New York: Teachers College Press.

Chapter 5
Understanding Objectivity in Research Reported in Reference Works

5.1 Evaluation of Research Reported in *International Handbook of Research in History, Philosophy and Science Teaching* (HPST)

5.1.1 Method

HPST is the first handbook (http://www.springer.com 978-94-007-7653-1) devoted to the field of historical and philosophical research in science and mathematics education. The handbook has 76 chapters written by 125 authors from 30 countries, which makes it truly an international endeavor. More than 300 reviewers from the disciplines of history, philosophy, education, psychology, mathematics, and natural science contributed with their expertise to its elaboration. In order to understand the rationale of the handbook it is important to consider the following invitation that was sent to the prospective authors of the different chapters:

> The guiding principle for the *Handbook* chapters is to review and document HPS [History and philosophy of science]-influenced scholarship in the specific field, to indicate any strength and weaknesses in the tradition of research, to draw some lessons from the history of this research tradition, and to suggest fruitful ways forward .... The expectation is that the handbook will demonstrate that HPS contributes significantly to the understanding and resolution of numerous theoretical, curricular and pedagogical questions and problems that arise in science and mathematics education. (Matthews, 2014b, p. 7)

This clearly shows the wide ranging and multiple objectives of the handbook that can provide guidance for future research as well as curricular and pedagogical feedback to those working in the educational field. Based on the subject index of

---

This chapter reports the evaluation of research reported in the following reference works: (a) *International Handbook of Research in History, Philosophy and Science Teaching* (Editor: M.R. Matthews, Springer, 2014a); and (b) *Encyclopedia of Science Education* (Editor: R. Gunstone, Springer, 2015).

© Springer International Publishing AG 2018
M. Niaz, *Evolving Nature of Objectivity in the History of Science and its Implications for Science Education*, Contemporary Trends and Issues in Science Education,
DOI 10.1007/978-3-319-67726-2_5

the handbook, I found eight chapters that discussed some aspect of objectivity. Following the guidelines based on Charmaz (2005), presented in Chap. 3, and in order to facilitate credibility, transferability, dependability, and confirmability (cf. Denzin & Lincoln, 2005) of the results, I adopted the following procedure: (a) All the eight chapters from the *International Handbook of Research in History, Philosophy and Science Teaching* (HPST) were evaluated and classified in one of the five levels (I–V); and (b) After a period of approximately 3 months all the articles were evaluated again and there was an agreement of 91% between the first and the second evaluation. It is important to note that the authors of these chapters were not necessarily writing about objectivity, but rather referred to it in the context of their selected topic (Appendix 5 provides a complete reference to each of these eight chapters that can provide readers with an overview of the topic of interest).

5.1.2 Results and Discussion

Each of the eight chapters in the Handbook was evaluated (Levels I–V) with respect to the context in which they referred to objectivity. Levels I–V are the same as those used in Chap. 3. Based on the treatment of the subject by the authors, following sections (categories) were developed to report and discuss the results. These sections are presented in alphabetical order. Distribution of the chapters according to the Level (for complete details see Appendix 6) was the following: Level I = 0; Level II = 4; Level III = 3; Level IV = 1; and Level V = 0. It is important to note that some of the chapters could have easily been placed in more than one section.

5.1.2.1 Cultural Studies and Objectivity

Cultural studies provide a critique of the traditional science education programs with the objective of a fundamental reconceptualization. Starting in 2006, these attempts have received additional support from the publication of the journal *Cultural Studies of Science Education*, CSSE (Springer). Based on a critical review of the literature, McCarthy (2014) explores the relationship between objectivity, sociology of science (R. Merton), and feminist studies (S. Harding), and then draws implications for the role of cultural studies in science education. According to Merton (1938), the ethos of science is characterized by universalism, communism, disinterestedness, and organized skepticism (see Chap. 1 for details). In science, universalism leads to objective sequences of verified knowledge that precludes particularism, and this led McCarthy (2014) to conclude: "Merton's universalism rests upon the objectivity of the world itself, the object of scientific inquiry" (p. 1928). Furthermore, Merton rejected the notion that scientific knowledge could have a particular cultural, national or class-based content. In contrast, Harding has argued that the claims of Western modern science to universality and objectivity should be rejected as illusions. Furthermore, she has suggested that all theories in natural science are social constructs and are strongly influenced by social and cultural factors. Harding's (1998) work has generally been endorsed by science educators who publish in CSSE. At this stage it is important to note McCarthy's (2014) following thought-provoking remarks:

> The problem here is that Harding's conclusion relies on an assumption that any degree of influence on science by social factors negates the claim of science to objectivity. But, absolute objectivity need not be conceptually required. A greater degree of objectivity is the value to be sought in scientific inquiry. The institutional structure and norms of scientific inquiry support the goal of achieving greater objectivity. (p. 1935)

It is important to note that the role of social, cultural, and other factors is important in cutting-edge research when the stakes are high and the results uncertain, namely "science in the making." Furthermore, it is reasonable to assert that "absolute objectivity" always remains a part of a never ending quest. In a similar vein, Machamer and Wolters (2004) have argued that objectivity comes in degrees. This also approximates to what Daston and Galison (2007) have referred to as the evolving nature of objectivity in the history of science. At this stage it is interesting to note that McCarthy concurs with Daston and Galison (2007) with respect to the rise of objectivity in the 1800s as an epistemic virtue associated with scientific inquiry. Furthermore, according to McCarthy (2014): "Prior to the adoption of the ideal of objectivity, they [Daston & Galison] claim, it was common practice for observers to discard discordant observations as defective and, guided by personal intuitions of truth and essential form, to seek out the examples that would seem to verify the favored theory" (p. 1952). Indeed, "discarding discordant observations" was common practice in the history of science, a period that Daston and Galison refer to as "truth to nature." This was followed by the ideal of mechanical objectivity, which in turn faced serious criticisms and was followed by "trained judgment." Classified as Level IV as it approximates to the changing/evolving nature of objectivity.

Interestingly, McCarthy considers that Harding's position has many tensions. For example, she accepts an orderly and objective reality but to develop a body of objective knowledge of reality (including beliefs that approach truth) is considered impossible. Perhaps a similar tension exists in the ideas of Roth (2008) who considers that scientific discourse need not be considered as superior to hybrid discourse based on familiar beliefs that originate outside the science classroom. At this stage it would be interesting to consider if Roth's ideas correspond to "truth to nature" or mechanical objectivity.

5.1.2.2 Feminism and Objectivity

Feminist critiques of science are generally classified as: (a) Feminist empiricism, which maintains that standard methods of science in themselves are good. Sexist science deviates from these existing canons of good science; (b) Feminist standpoint theory, which challenges standard account of science. Among others, it draws inspiration from Marx, as the worker has to understand his own viewpoint and that of the boss and this requires a struggle and consequently leads to a better understanding of how things are. Similarly, females have a better understanding than their male counterparts. Despite this claim, "… it remains wedded to the ideal of scientific objectivity" (Mackenzie, Good, & Brown, 2014, p. 1077); (c) Feminist postmodernism, which embraces a form of relativism that questions the standard view of science and is skeptical of objectivity. It emphasizes the "local" and ignores the inherent inconsistencies in "local narratives." Although, many feminists have considerable sympathy for post-modernism, the vast majority of feminist philosophers of science do not. Furthermore, most feminists emphasize the importance of social factors in the progress of science based on a diversity of views. Based on these considerations, Mackenzie et al. (2014) concluded:

> Feminist philosophers of science who insist on taking these sorts of social factors into account while still upholding scientific objectivity have probably improved the standard account of science considerably, especially as it applies to the social sciences. Their ranks include Anderson, Harding, Kourany, Longino, Nelson, Okruhlik, Wylie and many others. (p. 1078)

Furthermore, Mackenzie et al. (2014) recognize the importance of diversity of views and how that shows the problematic nature of objectivity: "But objectivity is undermined if the objective correctness of a claim is taken to be what is endorsed by a privileged point of view …. That privileging would leave no possibility for the chosen point of view to be itself mistaken. For objectivity to be possible, no point of view can be globally privileged. Objectivity consists in a perspectival form, rather than any possibility of a non-perspectival content" (p. 1066). This presentation approximates on the one hand to Giere's (2006a, b) perspectivism and at the same time also Daston and Galison's (2007) changing nature of objectivity. Classified as Level III.

5.1.2.3 Mathematics and Objectivity

There is some consensus that mathematical propositions are not empirically falsifiable, and thus possess the absolute certainty of analytical statements or logical truths. However, these are not immune to different forms of criticism, especially if they fail to solve a particular problem, and in this sense mathematics is not radically different from science. According to Glas (2014), teachers and students consider mathematics as one teaching subject in which practical experiences are irrelevant and where there are single right answers to all questions, whose correctness is beyond doubt (p. 731). In this context it is important to note that:

> Especially the most advanced sciences are very much like mathematics in that their conceptual apparatus and organisation are to a large extent non-observational and self-supporting. Only think of the theory of relativity, which depended on the nonempirical principle of relativity and the non-Euclidean geometries developed in the second half of

the nineteenth century. Einstein's achievements relied on thought-experiments and mathematics; empirical methods became relevant only when confirmation or corroboration was called for. (Glas, 2014, p. 732)

Recent research based on string theory has provided another example of how mathematical structures can provide insight with respect to the development of new ideas, and Dawid (2006) has expressed this in cogent terms:

... the fact that string theory has not been corroborated by any direct experimental evidence thus far seems to render it a mere theoretical speculation ... For many years now, the string community has been one of the largest communities in all of theoretical physics and has produced the majority of the field's top-cited papers ... The fact that an entirely unconfirmed speculative idea can assume such a prominent position in a mature scientific field is quite astonishing. (p. 299)

Similarly, Glas (2014) has endorsed the role played by string theory in scientific progress.

Lakatos' (1976) work on *Proofs and Refutations* can be regarded as the seminal text for the quasi-empirical approach to mathematics. Based on the work of Pólya and Popper, Lakatos developed his idea of *heuristics*, namely the use of counterexamples, suggesting falsification, as a critical tool for the achievement of growth of knowledge. Popper had not originally intended his methodology of conjectures and refutations to apply to mathematics and despite some differences endorsed Lakatos' initiative. According to Glas (2014) the objectivity of mathematics is inseparably linked with its criticizability, which was recognized by Popper (1981) in his *Objective Knowledge*, in the following terms: "Language becomes the indispensable medium of critical discussion. The objectivity, even of intuitionist mathematics, rests, as does that of all science, upon the criticizability of its arguments" (pp. 136–137). Furthermore, Popper tried to bring the (third) world of objective ideas down to earth by analyzing its relationships with the physical (first) and the mental (second) world (Glas, 2014, p. 746). This helps to overcome traditional dualisms such as between realism and constructivism. In other words, it is possible to be a constructivist and a realist at the same time with respect to the objective content of mathematics. According to Daston and Galison (2007), in the late nineteenth and early twentieth century, the proponents of structural objectivity were mostly mathematicians, physicists, and logicians who waged a war on images, and instead endorsed mathematical structures. Classified as Level III.

5.1.2.4 Multiculturalism and Objectivity

The importance of indigenous knowledge and wisdom and Western modern science and the ensuing conflict in the context of naturally occurring events in school science programs has been the subject of a study by Horsthemke and Yore (2014). They refer to the issues raised by Aikenhead's *cultural border crossing* (transition between a student's life-world and school science) and Jegede's *collateral learning* (cognitive conflicts arising from cultural differences between students' life-world and school science). These conflicts are expressed by Aikenhead and Jegede (1999) in a cogent and picturesque style:

> A simple example of collateral learning is illustrated by a rainbow. In the culture of Western science, students learn that the refraction of light rays by droplets of water causes rainbows; in some African cultures, a rainbow signifies a python crossing a river or the death of an important chief. Thus, for African students learning about rainbows in science means constructing a potentially conflicting schema in their long-term memory. Not only are the concepts different (refraction of light versus pythons crossing rivers), but the epistemology also differs ("causes" versus "signifies"). (p. 276)

Most readers at this stage will recall that in many cultures students are exposed to indigenous ways of thinking not only about the rainbow but many other phenomena and issues, of which perhaps the controversy between creationism and evolution is most wide spread (this is especially so in the USA, cf. Skoog, 2005). As science educators it is important for us to respect and understand local (indigenous) sources of knowledge. However, at the same time we also have the responsibility for finding appropriate teaching strategies for introducing explanations of phenomena based on scientific knowledge. In this context, it is important to note that Le Grange (2004) considers that all knowledge is local. Horsthemke and Yore (2004) go beyond by stating that: "While it makes sense to say that 'all knowledge systems have localness in common', they also share objectivity and *trans*localness" (p. 1777). Classified as Level II. The idea of "*trans*localness" is important as it may signify that local knowledge can transcend and lead to wider acceptance. In the context of the controversy surrounding evolution the following advice can help to go beyond local knowledge: "It is perfectly acceptable for those running creationist institutions to critique evolution and to try to persuade those visiting such institutions that the standard evolutionary account is wrong. But just as science teachers with no religious faith should respect students who have creationist views, so creationists should not misrepresent creationism as being in the scientific mainstream. It is not" (Reiss, 2014, p. 1658). This is the crux of the issue, as many creationists by espousing diversity in the science curriculum justify the inclusion of creationism.

5.1.2.5   Nature of Science and Objectivity

According to Reiss (2014), much of school science experimentation is Popperian. For example, when we see a rainbow it is hypothesized that white light is split up into light of different colors due to refraction. We accept this until some new evidence causes it to be falsified. Kuhn (1970) critiqued Popper as falsificationism holds only during periods of "normal science," when there are no competing paradigms or research programs (as suggested later by Lakatos). Next, Reiss (2014) refers to the characterization of science by Robert Merton and then concluded: "Allied to the notion of science being open-minded, disinterested, and impersonal is the notion of scientific objectivity. The data collected and perused by scientists must be objective in the sense that they should be independent of those doing the collecting (cf. Daston & Galison, 2007)—the idealized 'view from nowhere'" (pp. 1642–1643). It is important to note that this characterization would be considered by Daston and Galison (2007) as coming very close to "mechanical objectivity" which was the subject of considerable criticism and later led to the formulation of "trained judgment." Classified as Level II.

5.1.2.6   Optics and Objectivity

Galili (2014) presented a historical reconstruction of the different ideas put forward to understand optics knowledge. Heron of Alexandria demonstrated the rule: the light path presents the shortest trajectory between any two points. Fermat in the seventeenth century advocated the extreme temporal rather than spatial path of light. Measurements confirmed the law of refraction as the sine ratio of the angles of incidence and refraction. Descartes suggested an additional mechanism of light refraction based on an analogy between light and the motion of a ball being hit downward by a tennis racket at a water surface. Although all these ideas could be considered as objective knowledge, they also had a subjective part. In the nineteenth century, subjective speculations regarding light propagation were removed by Fresnel's introduction of wave interference as a tool to apply Huygen's principle. Based on this reconstruction, Galili (2014) concluded: "The four theories of light illustrate an area of optics knowledge in the third world. Though very different in validity, they share the property of objectivity, remaining human, that is, a subject for refinement and falsification" (p. 116, in Popper's framework the third world is distinguished from the real world, the first one and the personal world, the second one). This presentation helps to understand the evolving nature of objectivity (despite subjectivity in the different historical ideas) and how this can help students to understand a better picture of optics knowledge. Classified as Level III.

5.1.2.7   Philosophy of Science and Objectivity

Research in science education has recognized the importance of philosophy of science for science education. Despite this recognition, philosophy per se is accorded much lesser importance. Many science teachers wonder: what does philosophy have to do with science? This is understandable as most science teachers are overloaded with teaching and are primarily concerned with science content and assessments, as outlined in the curriculum. This is all the more difficult to understand as science content is invariably based on the development of models, theories, laws, and explanations. A student learning atomic structure in almost all parts of the world based on the atomic models of J.J. Thomson, E. Rutherford, N. Bohr and others can easily ask: why do models change? This is precisely, where the philosophy comes in and we are faced with the question: are the teachers prepared to respond to such questions? Despite these concerns, most science textbooks and curricula generally ignore the philosophical background of how science progresses. In this context, Schulz (2014) has concluded:

> Moreover, the scientific tradition (as an integral part of Enlightenment culture) based on rationality, objectivity, and skepticism, which teachers have inherited, is equally challenged by strands of pseudoscience, irrationality, and credulity of the times … How can teachers illustrate these differences, especially the distinction between valid and reliable knowledge from invalid ones (or natural from supernatural claims), without philosophical preparation? Yet it is not just the classroom, contemporary media discourse, or pop culture that is infused with questions, beliefs, claims, and counterclaims of philosophical significance, but like-wise the evolution of science itself. (p. 1268)

Indeed, science teachers need to go beyond the simple exposition of science content and instead explore the evolution of science itself, namely "science in the making" (cf. Niaz, 2012). Classified as Level II.

5.1.2.8   Research Methodology and Objectivity

Research methodology is important for understanding both educational research and its implementation in the science classroom. Taber (2014) differentiates the methodologies used in qualitative and quantitative research in the following terms:

> … the term quantitative research is also sometimes reserved for the use of hypothesis testing procedures, excluding studies that analyse quantitative data to offer purely descriptive statistics. Similarly, some authors limit the use of the term qualitative research to studies that admit the necessity of a subjective element … and are based on an interpretative approach that does not claim *objectivity in the normal scientific sense*—because it is argued that some kinds of social phenomena can only be understood through the intersubjectivity—and that the kind of *detached observer who could claim objectivity* would not be able to access suitable data for the study. (p. 1871, italics added). Classified as Level II.

This presentation differentiates quantitative from qualitative research as in the latter *objectivity in the normal scientific sense* cannot be claimed and the data is not collected by a *detached observer who could claim objectivity*. Daston and Galison (2007) have shown that even in the natural sciences (quantitative studies) at times objectivity in the normal scientific sense cannot be achieved nor the presence of a detached observer guaranteed. In such cases *mechanical objectivity fails and the scientific community has to resort to trained judgment*. Also see Campbell (1988a, b) on how objectivity in physical science depends on a social process of competitive cross-validation. This shows that the criterion of the type of data and how it is analyzed is not necessarily a good criterion for differentiating between qualitative and quantitative research. With respect to the key issue of the nature of teaching and learning, Taber (2014) stated:

> One common type of study compares learning in two "comparable" classes where teaching by an innovative ("progressive") approach is compared with teaching through a "traditional" approach. This immediately creates problems for making a fair comparison whether the teaching is carried out by the same teacher (will they be as equally adept and

enthusiastic in both conditions?) or different teachers (who inevitably will bring different skills, and knowledge to their teaching). (pp. 1875–1876, italics added)

Interestingly, if we accept these arguments based on the traditional conceptualization of objectivity, most research in science education would be considered as "flawed." What is at issue is not if we can have a fair comparison by asking the same teacher to provide progressive and traditional instruction or alternatively have different teachers for the two comparison groups. What is at issue is how we can ascertain that the teacher who provides progressive instruction is fully trained and "immersed" in the relevant literature. This is all the more important if the experimental treatment is based on some deep conceptual/philosophical issues that require considerable time effort and commitment to understand the dynamics of the classroom environment. Does this mean that a fair comparison is impossible? A fair comparison is possible, however with some constraints—just like objectivity comes in degrees so does validity in research comes in degrees (cf. Machamer & Wolters, 2004). In this context, Niaz (2011) has argued that, "Finally, if the problem precedes the method, and requires a historical reconstruction of the science topic to be introduced in the classroom based on how science is practiced by scientists, it is essential that the methodologists follow the practicing researchers" (p. 198). This is quite similar to what a scientist might have to do in the laboratory on finding that while doing cutting edge research, known and established experimental procedures have to be changed based on creative and innovative strategies. History of science can provide many examples of such episodes (cf. Niaz, 2016).

5.2   Evaluation of Research Reported in the Encyclopedia of Science Education

5.2.1   Method

Encyclopedia of Science Education (ESE) is the first encyclopedia published by Springer that relates to science education. It consists of a comprehensive set of entries listed alphabetically with considerable cross referencing. Among others the following broad areas of interest to science education are included: learning, teaching, curriculum, assessment and evaluation, science education in and out of school contexts, nature of science, socio-cultural dimensions, teacher education, and information technology. The purpose of the ESE is to record and represent work done, but in a way that establishes a research platform for the future, to lay out achievements while also highlighting continuing debates, issues to be addressed, and conundrums that require attention. Thus, ESE not only reports work that has already been done but also establishes a research agenda for the future. In November 2015, I made an online literature search on the website of ESE (http://refworks.springer.com/ScienceEducation) with the key word "objectivity" and found 12 entries that discussed some aspect of objectivity. Following the guidelines based on Charmaz (2005), presented in Chap. 3, and in order to facilitate credibility, transferability, dependability, and confirmability (cf. Denzin & Lincoln, 2005) of the results, I adopted the following procedure: (a) All the 12 entries from the *Encyclopedia of Science Education* (ESE) were evaluated and classified in one of the five levels (I–V, as outlined in Chap. 3); and (b) After a period of approximately three months all the articles were evaluated again and there was an agreement of 95% between the first and the second evaluation. It is important to note that the authors of these entries were not necessarily writing about objectivity, but rather referred to it in the context of their selected topic (Appendix 7 provides complete references to each of these 12 entries that can provide readers with an overview of the topic of interest).

5.2.2 *Results and Discussion*

Each of the 12 entries in the ESE was evaluated (Levels I–V) with respect to the context in which they referred to objectivity. Levels I–V are the same as those used in Chap. 3. Based on the treatment of the subject by the authors following sections (categories) were developed to report and discuss the results. These sections are presented in alphabetical order. Distribution of the entries according to the Level (for complete details see Appendix 8) was the following: Level I = 0; Level II = 6; Level III = 4; Level IV = 2; and Level V = 0. It is important to note that some of the entries could have easily been placed in more than one section.

5.2.2.1 Affect and Objectivity

Alsop (2015) has drawn attention to the need for recognizing affect as an important factor in teaching and learning science. Following are some of the salient aspects of this perspective: (a) Affect has considerable influence over what happens in the classroom; (b) Some emotions (such as happiness, pleasure, delight, thrill, and zeal) act to potentially enhance learning and optimize student achievement, while other emotions (such as boredom, sadness, distress, regret, gloom, and grief) can lead to lack of concentration, curiosity and insight and thus do not facilitate learning; (c) The affective and emotional encounters and relationships that we develop within pedagogies and with knowledge are profoundly and deeply important; (d) Studies of affect in science education are theoretically wide ranging and empirically diverse. Researchers, for instance, have worked on particular motivational constructs including emotional intelligence, self-efficacy, task value, and achievement goals; (e) A major concern of science education is cognition, and conceptual performance is highly rewarded. Despite this, there is a clear evidence base that affect and cognition are inseparable and mutually constitutive; and (f) Traditionally, students' difficulties are associated with conceptual demands and emotional demands are generally ignored. These considerations led Alsop (2015) to conclude:

There is a long associated history, of course, in which affect is framed as mainly undesirable, as a potential obstacle to enlightened, objective thought (especially in science). In departing from this history and holding onto the importance of affect, we open up profound questions of objectivity and subjectivities, questions that more often than not accompany popular Western narratives of mind and body duality. There are legitimate arguments that such a departure leads one to a history of science that is more consistent with the practices of sciences than history often seeks to represent (p. 20). Classified as Level IV.

This presentation raises important issues by recognizing the importance of affect which leads to recognizing the problematic nature of objectivity in scientific progress. Furthermore, this departure facilitates an understanding of the history of science that is more in consonance with actual scientific practice and hence the need for understanding "science in the making."

5.2.2.2   Bildung and Objectivity

Bildung is an important concept in German speaking and North European countries. In this regional and linguistic context it is the central notion describing the process of personal development and its outcome. *Bildung* is more than education, and there is no English term that denotes the concept accurately. Elaboration of the concept of *Bildung* has drawn inspiration, among others from Kant's ideas on Enlightenment, and Von Humboldt's idea that the individual and humanity are two facets that are strongly interrelated. Interestingly, the natural sciences were not considered as a domain contributing to *Bildung* and this generated considerable controversy, so much so that it took almost 100 years for the natural sciences to be integrated into the school curriculum of the higher educational institutions (e.g., the Gymnasium). W. Klafki advocated a modern understanding by emphasizing the relationship between *Bildung* and society based on self-determination, responsibility, reason, and solidarity. Lest this be considered as an individualistic conception he also emphasizes a second group of factors that include: humanity, humankind, humaneness, and objectivity. Given the difficulties involved in understanding *Bildung* one German scholar, Tenorth even suggested that *Bildung* can be regarded as a German myth, a pedagogical program, a political slogan, and an ideology of bourgeoisie. There has also been attempt to understand *Bildung* as "scientific literacy" (especially in the context of the PISA Project). However, Fischler (2015) considers that *Bildung* cannot be considered as the European version of scientific literacy. It is plausible to suggest that objectivity constitutes a part of *Bildung* as an academic objective (Classified as Level II).

5.2.2.3   Broadcast Media and Objectivity

Robinson (2015) traces the origins of broadcast media in radio and television and draws implications for science education (both formal and informal). For example, the purpose of the BBC, from the very beginning, was to "educate, inform and entertain." Technological advances, especially the Internet has provided new channels of communication. CERN, the European Organization for Nuclear Research, has a channel CERNNews that provides regular programs about the physics experiments conducted at their laboratories. Similarly, NASA TV has live coverage of its space research, viewable on computers and mobile devices. The Open universities in many countries have science programs targeted at particular audiences with the possibility of taking course credit. Recent developments show that even the commercial breaks during the science programs can include science content, which led Robinson (2015) to conclude: "It has been suggested that advertisements could be used to learn about science and, by examining them in detail, demonstrate the need for objectivity" (p. 138). It is plausible to suggest that broadcast media could also be helpful in discussing the details of important historical episodes in order to provide a better perspective of the role played by objectivity (especially trained judgment). Classified as Level II.

5.2.2.4 Constructivism and Objectivity

According to Taylor (2015) there are many versions of constructivism in the literature, with labels such as cognitive, personal, social, radical, cultural, trivial, pedagogical, academic, contextual, and ecological. Cognitive constructivism is based on J. Piaget's ideas of an active "constructing" mind of the individual student which had been largely overlooked by the dominant behaviorist teaching method of lecturing to silent classrooms. Personal constructivism based on the work of R. Driver, D. Ausubel, and G. Kelly discovered that students' intuitive understandings of their experiences are so strongly held that in many cases they block development of counterintuitive scientific concepts. Ernst von Glasersfeld's radical constructivism was promoted by science educators who were dissatisfied with the objectivism of personal constructivist pedagogy, where objectivism regards scientific knowledge as an accurate depiction of physical reality. Radical constructivism draws on Piaget's genetic epistemology which emphasizes the inherent uncertainty of the constructed knowledge of the world by all cognizing beings, including children and scientists. Some critics consider this stance of radical constructivism as approximating to relativism. Social constructivism draws on theories of social psychology of J. Lave and E. Wenger, which recognizes that students construct meaningful knowledge in communities of practice, similar to how scientists negotiate among peers. Furthermore, it draws on the social activity theory of L. Vygotsky, which emphasizes the development of language and thought. Critical constructivism was promoted by science educators sensitive to issues of social justice, and is based on the ideas of J. Kincheloe, P. Berger, T. Luckmann, and P. Freire's pedagogy of the oppressed. This form of constructivism considers the traditional science curriculum as based on oppressive ideologies that lurk like Trojan horses. Furthermore, critical constructivism is based within a cultural context that recognizes that Western modern science owes much to earlier developments in Africa, China, Japan, India, Persia, and Arabia. Integral constructivism is based on a dialectical perspective in which the different forms of constructivism (even including behaviorism) can be integrated into an ever-expanding repertoire. This mosaic of the different forms of constructivism and the inherent controversies can be confusing to a beginner. According to Niaz et al. (2003) this represents a continual critical appraisal of our theoretical formulations and shows the tentative nature of both science and educational theory, of which constructivism is an example. This perspective also provides an overview of how well science education enables students to understand and differentiate the epistemological status of scientific concepts, theories, and laws. Finally, Taylor (2015) concluded:

> For radical constructivism, the cornerstone concept of "objectivity" is reconceptualized as consensual agreement by scientific communities of practice. This instrumentalist perspective on knowledge production and legitimation is in close accord with David Bloor's "strong program" of the sociology of science knowledge (SSK) and with the philosophy of science of Thomas Kuhn who argued persuasively that scientific knowledge is "paradigm bound." (p. 220)

This conceptualization of objectivity as consensual agreement within scientific communities of practice comes quite close to what Daston and Galison (2007) have referred to as "trained judgment." Classified as Level IV.

5.2.2.5 Curriculum, Values and Objectivity

Besides subject matter content, a curriculum generally expresses the purposes, goals, and aims of education. Science content in itself is associated with the values that guide scientific research. A scientist needs to be curious enough to explore the subject of his inquiry. The purpose of the study helps in generating data and its analysis leads to patterns and regularities. This process of interpreting the data means that the scientist is using his/her particular form of reasoning and understanding, leading to the utilization of some data to a greater extent, especially the one that helps in the elaboration of a model or explanation. This inevitably leads the scientist to prefer his own meaning, based on his values while trying to understand the data. The next step is to communicate these findings, based on her/his values and interpretations to the scientific community in order to seek criticisms and eventually achieve consensus. Based on these considerations Corrigan (2015) suggests that science education needs to communicate this perspective of the scientific endeavor to students and concluded that:

> It has been quite a common notion amongst some scientists, science educators, and the general community that science is "value-free." Such ideas have often been perpetuated in the study of science and in science communication, particularly through the focus of science being objective. But objectivity is not really possible as science is a human construction—a way of explaining our natural world. (Corrigan, 2015, p. 256)

This presentation based on "how science is done" helps to understand that it cannot be value-free and thus facilitates an understanding of the problematic nature of objectivity and its importance for science education. Classified as Level III.

5.2.2.6   Ethnoscience and Objectivity

Stewart (2015) considers the concept of ethnoscience as vexatious as it literally means "cultural science," which flouts the criteria according to which science is culture-free:

> The concept of ethnoscience reminds science of its basis in culture: as a human form of knowledge, science can aspire to, but never fully attain, the criteria for knowledge that are regarded as its essential characteristics, such as objectivity and universality. This humbling of science is the actual value for science of the vexatious concept of ethnoscience (p. 401). Classified as Level III.

Ethnoscience refers to various forms of indigenous or native science, such as "African science" or "Maori science" (New Zealand). Maori science educators have protested against the conflation of the terms "Western science" and "science," which leads to Eurocentrism and elitism in the secondary school science curriculum and consequently alienate almost all Maori students. Stewart has argued that what is at issue is not the recognition of "Maori science" as an alternative form of science but the fact that some of the essential characteristics of science such as objectivity are difficult to attain, and ethnoscience provides an opportunity to understand the limitations of science. In a similar vein, Aikenhead and Michell (2011) have suggested that in Eurocentric science during the peer-review process scientists scrutinize their ideas, methods, data, and arguments in order to achieve some degree of consensus, which helps them to approximate the ideal of objectivity. Furthermore, these authors while recognizing that objectivity remained a powerful and useful ideal concluded "The public storyline that scientists attain objectivity is a myth …. The ideal of objectivity fails in the reality of practice" (p. 41).

5.2.2.7   Gender and Objectivity

The education of girls and women has been neglected in most cultures and generations. An organizing principle in most societies is that those practices and attributes associated as masculine are more highly valued than those associated as feminine. However, in recent decades substantial gains have been reported with respect to equal opportunities for women. The equitable participation of men and women in science careers is based on the concern that everyone needs to engage actively in science-related issues in their everyday lives. Despite the gains in some fields (e.g., computer science, economics, and physics) women are still underrepresented. Sciences are in general regarded as more masculine than other pursuits. Based on these considerations, Brickhouse (2015) has concluded:

> The masculine association of the sciences is related not only to male/female participation but also to its epistemology …. Scientific knowledge, like other forms of knowledge, is gendered. Culturally defined values associated with masculinity (objectivity, reason) are also those values most closely aligned with science. While this association of masculine

values with science dramatically oversimplifies the practice of science, it is nevertheless a powerful fiction that may serve to exclude those who do not hold to these values. (p. 441)

This helps to understand a complex issue, namely in most cultures masculinity is associated with objectivity and reason, and these are precisely the values that are also traditionally associated with science. Furthermore, the role played by objectivity in the practice of science is problematic and in itself needs a reconceptualization (Daston & Galison, 2007). Classified as Level III.

According to Scantlebury (2015) gender is a social category and as such influences social interactions, including schooling and science education. Furthermore, as a social category, gender is constituted on the structural, the symbolic, and the individual levels in society. The *structural level* examines the division of labor by gender. In science, there is a consistent pattern of more women working in the biological sciences compared to the physical sciences. The "gendered" explanation for this is that the former is more feminine and the latter masculine. One example is women's involvement with ecological feminism to improve living conditions of their families. Science educators generally use gender at the *individual level* rather than in a social context. In other words science education research does not offer a critique of the gender concept but rather focus on comparing female and male students' achievement, participation, and attitude toward science. Finally, Scantlebury (2015) elaborated on the *symbolic level* in the following terms:

The symbolic level of gender uses dichotomies where the oppositional pairs are assigned a feminine and masculine meaning (e.g., nature/culture, emotion/rationality, subjectivity/objectivity) that infers what are appropriate practices for women and men. For example, the symbolic level describes science as rational, difficult, and hard, with disembodied knowledge. Thus, both structurally and symbolically, science is a masculine gender practice. In contrast, teaching, especially children, is described as nurturing and caring, which is symbolically feminine. (p. 984)

This presentation associates objectivity with masculine gender practice. However, it does not discuss the problematic nature of objectivity in such a context. Classified as Level II.

5.2.2.8 Religious Education and Objectivity

Reiss (2015) discusses the nature of religion and the nature of science and then examines how religion and science might relate. The aims of religious education and science education are considered and ways of teaching science to take account of religious beliefs are examined. Role played by objectivity in the construction of scientific knowledge is acknowledged in the following terms:

… while the subject matter of science has varied considerably over the centuries, often because of advances in instrumentation, its principal approaches to building up reliable knowledge are fairly consistent. Of central importance is the objectivity of such knowledge—i.e., the knowledge should be independent of the person generating it (unlike, e.g., the work of painters, of composers and novelists, and perhaps of psychoanalysts)—and, relatedly, that such knowledge can be rigorously tested, often by experiment, though such experimentation is less direct for the historical sciences, such as much of geology and evolutionary biology, and for certain other sciences, e.g., astronomy. (Reiss, 2015, p. 832)

This presentation raises important issues especially with respect to how science develops while in the process of being established and the work of painters and novelists. Leon Cooper (Nobel Laureate in physics) has drawn an analogy between a style of painting (impressionism) and scientific progress: "I believe, in some ways, the scientist can be compared to the painter. The impressionists, for example, were accused of not being able to see things as they are. But, having imposed their way of viewing—their vision of the world—it has become a cliché now to see things as the impressionists did" (Reproduced in Niaz, Klassen, McMillan, & Metz, 2010b, p. 48). Another Nobel Laureate in literature, Mario Vargas Llosa (2010) has declared: "Literature is a false representation of life that nevertheless helps us to understand life better, to orient ourselves in the labyrinth where we are born, pass by, and die." This clearly shows how the work of the scientist and the novelist may be considered as "false" in the beginning and over time the community comes to recognize its merits. In this context understanding objectivity is important as both the novelist and the scientist are always looking for "something" that they do not know how to find and still keep persevering in this never-ending quest for yet another stepping stone (for more details see Niaz, 2016, Chap. 3). Reiss (2015, p. 833) refers to a similar experience faced by biology and earth science instructors while teaching evolution. Despite the consensus with respect to the difficulties involved in teaching evolution, it still remains a controversial subject for classroom teachers. Classified as Level II.

5.2.2.9 Science Studies and Objectivity

The origin of the science studies program can be traced to the work of Robert Merton (1910–2003) who articulated a view of the social and cultural norms of science, and Ludwig Fleck (1896–1996). Later the work of Thomas Kuhn (1922–1996) provided the catalyst that gave rise to science studies as an identifiable field in the 1960s. Rudolph (2015) then provides an account of subsequent changes in this field and its implications:

> The most famous of these was the science studies program at Edinburgh University. Scholars in this program developed what came to be called the sociology of scientific knowledge (SSK) approach that called into question the authority and objectivity of science. Taking their cue from Kuhn's assertion in *Structure* that revolutionary changes in science occur by means other than rational consideration of empirical evidence, Edinburgh scholars such as Barry Barnes and David Bloor argued that the emergence of scientific theories is significantly influenced by the social and cultural commitments to which scientists adhere. (p. 915)

This presentation refers to how SSK questioned the authority and objectivity of science but does not go beyond and refer to the problematic nature of objectivity. Classified as Level II.

5.2.2.10   Values and Objectivity

In the context of values and Western science, Irzik (2015) discusses the issue of whether science is value-free or not and then goes on to note that the job of the scientists is to discover facts about the world and not to pass any value judgments. According to Irzik (who traces the origin of value-free science to Bacon and Galileo), this perspective is supported by Mertonian norms of universalism, organized skepticism and disinterestedness and besides this, non-epistemic values such as: scientists should not fabricate, distort, or suppress data. Consequently, scientific theories and claims should be accepted or rejected on empirical grounds, not on social, political, moral, and religious considerations. These considerations led Irzik (2015) to conclude:

> In the light of the distinction between epistemic and non-epistemic values, the doctrine of value-free science can be formulated more accurately: non-epistemic values should play no role in scientific inquiry; any "outside" interference with the workings of science has devastating effects on scientific progress and objectivity, as exemplified by the Galileo and Lysenko affairs. (p. 1093)

Outside interference would be considered to be detrimental to scientific progress by most observers. In the twentieth century besides Lysenko we also have the examples of L. Pauling and A. Sakharov, who were ostracized for their views on the development of nuclear weapons. Irzik (2015) also recognizes the role played by some non-epistemic values such as ethical, social, and economic: "In short, not all influences of non-epistemic values on science are necessarily damaging to the progress, reliability, and objectivity of science" (p. 1094) and that science is, to varying degrees, value-laden in its social organization, language, methods, and even theories. These considerations lead Irzik to ask a thought provoking question: "If the ideal of value-free inquiry is flawed, what is to replace it?" (p. 1095). Furthermore, history of science and the historical evolution of objectivity show that the ideal of value-free science is at best a chimera. Based on Longino (2002), Irzik suggests that a promising alternative is "social value management," which incorporates non-epistemic values into science, provided they are all publicly subjected to rigorous critical scrutiny by taking into account all perspectives. (Classified as Level III).

Values are also considered a critical aspect of science education and can have a major influence on a student's behavior and attitude. Furthermore, curricula in science education not only emphasize the need to produce scientists but also empower all citizens with knowledge, skills, and values needed in a technological society. Based on these considerations Cavas (2015) considers that values in science are not so different from values in general, such as: "… valuing objectivity, accuracy, precision, pursuit of truth, and problem-solving; valuing human significance, the protection of human life, and balancing safety and risk; valuing intellectual honesty and academic honesty; valuing courage and humility; and valuing decision making and willingness to suspend judgment" (p. 1090). This is a fairly standard form of recognizing values in science education and of course "objectivity" forms an important part of such presentations (Classified as Level II). Nevertheless, the inclusion of "willingness to suspend judgment" is not included so frequently and thus provides an opportunity to discuss thought-provoking issues. Holton (1978a) refers to the importance of "willingness to suspend judgment" in his historical reconstruction of the oil drop experiment.

This chapter provides examples of research reported in reference works (HPST and ESE) that facilitate a wide range of perspectives with respect to objectivity. Conclusions based on these findings will be integrated with those from other chapters and presented in Chap. 7.

References

Aikenhead, G., & Jegede, O. J. (1999). Cross-cultural science education: a cognitive explanation of a cultural phenomenon. *Journal of Research in Science Teaching, 36*(3), 269–287.

Aikenhead, G., & Michell, H. (2011). *Bridging cultures: indigenous and scientific ways of knowing nature*. Toronto: Pearson Education Canada.

Charmaz, K. (2005). Grounded theory in the 21st century: applications for advancing social justice studies. In N.K. Denzin & Y.S. Lincoln (Eds.), *The Sage handbook of qualitative research* (3rd ed., pp. 507–535). Thousand Oaks, CA: Sage Publications.

Daston, L., & Galison, P. (2007). *Objectivity*. New York: Zone Books.

Dawid, R. (2006). Underdetermination and theory succession from the perspective of string theory. *Philosophy of Science, 73*, 298–322.

Denzin, N. K., & Lincoln, Y. S. (2005). Introduction: the discipline and practice of qualitative research. In N. K. Denzin & Y. S. Lincoln (Eds.), *The Sage handbook of qualitative research*. 3rd ed (pp. 1–32). Thousand Oaks: Sage.

Giere, R. N. (2006a). Perspectival pluralism. In S. H. Kellert, H. E. Longino & C. K. Waters (Eds.), *Scientific pluralism* (pp. 26–41). Minneapolis: University of Minnesota Press.

Giere, R. N. (2006b). *Scientific perspectivism*. Chicago: University of Chicago Press.

Harding, S. (1998). *Is science multi-cultural? Postcolonialisms, feminisms, and epistemologies*. Indianapolis: Indiana University Press.

Holton, G. (1978a). Subelectrons, presuppositions, and the Millikan-Ehrenhaft dispute. *Historical Studies in the Physical Sciences, 9*, 161–224.

Irzik, G. (2015). Values and Western science knowledge. In R. Gunstone (Ed.), *Encyclopedia of science education* (pp. 1093–1096). Heidelberg: Springer.

Kuhn, T. (1970). *The structure of scientific revolutions*. 2nd ed Chicago: University of Chicago Press.

Le Grange, L. (2004). Western science and indigenous knowledge: competing perspectives or complementary frameworks? *South African Journal of Higher Education, 18*(3), 82–91.

Longino, H.E. (2002). *The fate of knowledge*. Princeton, NJ: Princeton University Press.

Machamer, P., & Wolters, G. (2004). Introduction: science, values and objectivity. In P. Machamer & G. Wolters (Eds.), *Science, values and objectivity* (pp. 1–13). Pittsburgh: University of Pittsburgh Press.

Matthews, M. R. Ed. (2014a). *International handbook of research in history, philosophy and science teaching*. Dordrecht: Springer.

Matthews, M. R. (2014b). Introduction: the history, purpose and content of the Springer *International Handbook of Research in History, Philosophy and Science Teaching*. In M. R. Matthews (Ed.), *International handbook of research in history, philosophy and science teaching* (pp. 1–15). Dordrecht: Springer. vol. I.

McCarthy, C.L. (2014). Cultural studies in science education: philosophical considerations. In M.R. Matthews (Ed.), *International handbook of research in history, philosophy and science teaching* (Vol. III, pp. 1927–1964). Dordrecht: Springer.

Niaz, M. (2011). *Innovating science teacher education: a history and philosophy of science perspective.* New York: Routledge.

Niaz, M. (2012). *From 'Science in the Making' to understanding the nature of science: an overview for science educators.* New York: Routledge.

Niaz, M. (2016). *Chemistry education and contributions from history and philosophy of science.* Dordrecht: Springer.

Niaz, M., Abd-El-Khalick, F., Benarroch, A., Cardellini, L., Laburú, C. E., Marín, N., & Tsaparlis, G. (2003). Constructivism: defense or a continual critical appraisal—A response to Gil-Pérez, et al. *Science & Education, 12,* 787–797.

Niaz, M., Klassen, S., McMillan, B., & Metz, D. (2010b). Leon Cooper's perspective on teaching science: an interview study. *Science & Education, 19,* 39–54.

Popper, K. R. (1981). *Objective knowledge: an evolutionary approach.* Oxford: Clarendon.

Reiss, M. J. (2014). What significance does Christianity have for science education? In M. R. Matthews (Ed.), *International handbook of research in history, philosophy and science teaching* (pp. 1637–1662). Dordrecht: Springer.

Roth, W.-M. (2008). Bricolage, metissage, hybridity, heterogeneity, diaspora: Concepts for thinking science education in the 21st century. *Cultural Studies of Science Education, 3,* 891–916.

Skoog, G. (2005). The coverage of human evolution in high school biology textbooks in the 20th century and in current state science standards. *Science & Education, 14*(3–5), 395–422.

Vargas Llosa, M. (2010). Nobel Prize in Literature acceptance speech. http://www.nobel-prize. org/nobel_prizes/literature/laureates/2010/vargas_llosa-lecture. Accessed 11 December 2010.

Chapter 6
Science at a Crossroads: Transgression Versus Objectivity

6.1 Introduction

Chapter 1 dealt with the historical sequence in the work of Daston and Galison (2007), based on mechanical objectivity as a reaction to truth-to-nature and later trained judgment as a reaction to both the previous forms of scientific judgment (objectivity). In this chapter, I first explore the relationship between transgression and objectivity and then study the importance of Scanning tunneling microscope (STM) and the Atomic force microscope (AFM) for chemical research (nanotechnology) and how these are presented in general chemistry textbooks. STM was invented by Gerd Binnig and Heinrich Rohrer of IBM's Zurich Research Laboratory in 1981. Five years later both shared the Physics Nobel Prize for their invention. Atomic Force Microscope (AFM) was introduced in 1989 to better image nonconducting samples. It is now the most widely used scanning probe microscope. It works in gases or liquids. This is particularly helpful in studying biological specimens and biochemical processes in their natural environment. No staining is needed and it avoids any damage due to high-energy radiation. In order to facilitate a perspective of objectivity related to scientific (chemical) research and general chemistry textbooks, on February 20, 2016, I sent the document entitled "Transgression and Objectivity" (reproduced below)—to Roald Hoffmann, Nobel Laureate in chemistry and active researcher in various fields related to science in general, philosophy of science and chemistry in particular.

6.2 Transgression and Objectivity

6.2.1 Introduction

Note: At the end of the Introduction, questions for Professor Hoffmann are presented.

Although objectivity is not synonymous with truth or certainty, it has eclipsed other epistemic virtues and to be objective is often used as a synonym for scientific

© Springer International Publishing AG 2018
M. Niaz, *Evolving Nature of Objectivity in the History of Science and its Implications for Science Education*, Contemporary Trends and Issues in Science Education,
DOI 10.1007/978-3-319-67726-2_6

in both science and science education. Daston and Galison (2007) have constructed the evolving nature of scientific judgment through the following historical phases: truth-to-nature (eighteenth century), mechanical objectivity (nineteenth century), structural objectivity (late nineteenth century), and finally trained judgment (twentieth century). Each of these regimes did not supplant the other but they can coexist and supplement each other at the same time. The evolving nature of objectivity is important for science education as school and college science generally simplify complex historical episodes under the rubric of objectivity without really understanding that the underlying issues are dependent primarily on trained judgment. Followers of truth-to-nature were not particularly concerned if their images were objective. Those who followed mechanical objectivity reasoned that only objective images can avoid distortions. Followers of trained judgment frankly acknowledged the role played by subjectivity in their images. Actually, subjectivity is not a weakness of the self to be corrected or controlled, but rather it is the self (Daston & Galison, 2007, p. 374). The self, captured by subjectivity is highly individualized, whereas in some sense objectivity tries to eliminate individual peculiarities (p. 379). By the end of the twentieth century, this landscape started to change as, "For many scientists pursuing nanotechnology, the aim was not simply to get the images right but also to manipulate the images as one aspect of producing new kinds of atom-sized devices" (Daston & Galison, 2007, p. 282). For early twenty-first century nanoscientists (in domains such as fluid dynamics, particle physics, and astronomy), the issues that were important for those who wrestled with mechanical objectivity and trained judgment gave way to *images-as-tools*, namely images were meant to engineer things (p. 385). Scanning tunneling microscope (STM) and atomic force microscope (AFM) were two of the major techniques that helped to develop nanotechnology. Nano-manipulative atlases aimed not at depicting accurately that which naturally exists, but rather showing how nano-scale entities can be made, remade, cut, crossed, or activated (p. 391). Such images are examples of right depiction—of objects that are being made and not found. Daston and Galison (2007, p. 413) conceptualize right depiction as consisting of: (a) Representation (fidelity to nature), and (b) Presentation (fusing artifactual and natural). Representation has a long history that was variously understood as truth-to-nature (eighteenth century), mechanical objectivity (nineteenth century), and trained judgment (twentieth century). On the other hand presentation grew with nanotechnology in the late twentieth century and espouses object manipulation and aesthetics.

Interestingly, philosopher of science Ian Hacking (1983) had referred to this dilemma in cogent terms:

Maybe there are two quite distinct mythical origins of the idea of "reality." One is the reality of representation, the other, the idea of what affects us and what we can affect. Scientific realism is commonly discussed under the heading of representation. Let us now discuss it under the heading of intervention .... We shall count as real what we can use to intervene in the world to affect something else, or what the world can use to affect us. Reality as intervention does not even begin to mesh with reality as representation until modern science. Natural science since the seventeenth century has been the adventure of the interlocking of representing and intervening. It is time that philosophy caught up to three centuries of our own past. (p. 146)

Hacking concluded that the Baconian goal was to count as real that which can be used in the world through experimental intervention. For example, if you can spray positrons, these are real, as these can be used. Daston and Galison (2007) have recognized Hacking's foresight in the following terms:

> As Hacking saw it, in the early 1980s the long history of scientific depiction—tracing, drawing, sketching, even photographing—was doomed to fail. It would always be possible to invent a plausible reason to treat the reality of objects as merely a useful assumption, a helpful fiction. Hacking, seconding Bacon, contended that only *use* could provide a robust realism. It was a strong salvo in a long-standing debate over whether and under what conditions scientific objects may be taken as real. On the side of representation: we should take as real that which offers the best explanations. On the side of intervention: we should accept as real that which is efficacious. (p. 392, original italics)

This transition from representation to presentation (intervention in Hacking's terminology), although still in its early phases, can provide new opportunities to the researcher and also the institutional structure of research (p. 415). In other words, nanotechnology is not concerned about errors in our knowledge, nor if we are dealing with real objects but rather with creating and manipulating to construct a new world of atom-sized objects (p. 415). In this context, it is plausible to suggest that starting in the late twentieth century progress in science is at a crossroads. This is particularly important for science educators as on the one hand they have to study, depict, and explain what actually exists (representation) and at the same time be aware/explore the possibilities of what can be manipulated (presentation) to produce and create new possibilities and products. Nanotechnology can provide new materials such as biosensors that monitor and even repair bodily processes, microscopic computers, artificial bones, and lightweight strong materials.

Interestingly, Roald Hoffmann (2012), Nobel Laureate in chemistry, has also recognized Hacking's (1983) contribution in facilitating our understanding of the difference between representing and intervening:

> Is intervention an absolute correlate of experimental science? .... Surprisingly (given Bacon's startling metaphors and the resistance to them), it is only recently that philosophers have begun to explore the subject. An important examination of the question is found in Ian Hacking's idiosyncratic *Representing and Intervening*. In an important essay on "Experimentation and Scientific Realism," Hacking (1984) writes: "Interference and interaction are the stuff of reality." J.E. Tiles (1994), in an essay, "Experiment as Intervention," says of Hacking's analysis that it was "received in some quarters with a mixture of incomprehension and hostility." This is a minor surprise to us. We agree with Tiles that philosophers of science ought to consider that experiment does not only follow from observation (as both Aristotelians and positivist classical empiricists mistakenly and narrowly assume ...). At times, experiment is driven by an overtly interventionist stance. (Hoffmann, 2012, p. 124)

Actually, Hoffmann (2012) goes beyond and quotes the following passage about scientific realism from Hacking's (1983) idiosyncratic book:

> Experimental work provides the strongest evidence for scientific realism. This is not because we test hypotheses about entities. It is because entities that in principle cannot be "observed" are regularly manipulated to produce a new phenomena and to investigate other aspects of nature. They are tools, instruments not for thinking but for doing. (p. 262)

Indeed, Hacking's claim that experiments are done not to provide evidence for hypotheses but to produce new phenomena must have sounded incomprehensible (and difficult to understand) in 1983, not only to philosophers of science but also science educators.

According to Hoffmann (2012), an important characteristic of modern chemistry is that chemists inextricably mix macroscopic and microscopic viewpoints of substances and molecules in the productive work of their science (p. 35). Hoffmann refers to this practice as "violating categories" and then reflects in the following terms:

> I'm torn about this. I started out in chemistry perturbed by the mixing of categories around me, drunk on logic, mathematics, and symmetry. I was looking, as Primo Levi once was, for the theorems of chemistry. Eventually I came to peace with the multivalency of piecewise understanding around me. And I saw that partially *irrational reasoning* (oh, prettified for publication) led to stunning molecules and reactions. My perception of human beings, not just chemists, is

(a) That in the service of either creation or utility, they will naturally and deliberately violate all categorizations (here chemists inextricably mixing up molecules and compounds), and

(b) That the process of creation of the new depends essentially on the transgression of categorization.

> Point (a) is weak, and ultimately unimportant: people are people. Point (b) is stronger, with implications for philosophy: I want to claim that people are unlikely to make the new (art, science, religion, new people) without violating categories. I am here beyond philosophical holism, beyond intellectual bricolage, close to Feyerabend's prescription for "epistemological anarchism." Must be what too much poetry does. (Hoffmann, 2012, p. 36, italics added)

It is interesting to note that in his efforts to understand progress in chemistry, Hoffmann started by looking for the "theorems of chemistry" and ended up endorsing "transgression of categorization." Furthermore, the ideas of Hoffmann (2012) and Daston and Galison (2007) resemble to a great degree and following are some of the salient points of this resemblance:

(a) Both recognize Hacking's (1983) contribution in understanding scientific realism which led to the differentiation between "representation" and "intervention."

(b) Both acknowledge Hacking's contribution in the context of Francis Bacon's, contention that only "use" could provide a robust realism.

(c) Both question the role played by objectivity in the traditional scientific method, and how scientists have gone beyond the traditional understanding of progress in science.

Progress in science seems to be at the crossroads, precisely because images of scientific objects have surrendered any residual claim to being a version of "seeing," in a classical sense. We have come a long way when truth-to-nature strived to an idealized world, mechanical objectivity struggled to impose some form of blind sight and trained judgment approximated to bridges that facilitated the underlying essence of the scientific objects (e.g., the oil drop experiment).

References

(These formed part of the document "Transgression and Objectivity" sent to R. Hoffmann)

Daston, L., & Galison, P. (2007). *Objectivity*. New York: Zone Books.

Hacking, I. (1983). *Representing and Intervening*. Cambridge: Cambridge University Press.

Hacking, I. (1984). Experimentation and scientific realism. In J. Leplin (Ed.), *Scientific Realism*. Berkeley, CA: University of California Press.

Hoffmann, R. (2012). In J. Kovac & M. Weisberg (Eds.), *Roald Hoffmann on the Philosophy, Art, and Science of Chemistry*. New York: Oxford University Press.

Tiles, J.E. (1994). Experiment as intervention. *British Journal for the Philosophy of Science, 44*(3), 463-475.

6.2.2   Questions for Professor Hoffmann

Note: Following four questions were sent to Professor Hoffmann as part of the document "Transgression and Objectivity." To facilitate understanding, each of the questions is followed by the response (in italics) provided by Professor Hoffmann, in an email sent to me on February 23, 2016:

1. In a section entitled "Violating Categories" (Hoffmann, 2012, p. 35) you refer to "transgression of categorization" (p. 36). Would you agree that this transgression (irrational reasoning) in some sense approximates to what Daston and Galison (2007) have referred to as violating the rules dictated by objectivity?

   Yes, I agree. It's remarkable that you found the connection between these views, which came out at around the same time, but Daston and Galison's argument is much more soundly based in philosophy and aesthetics than mine. Yet we both refer to Hacking! (Hoffmann, R., Email to author, February 23, 2016a)

My comments: I am very pleased that Professor Hoffmann agrees with my interpretation that "transgression of categorization" approximates to the violation of rules dictated by objectivity as interpreted in the framework developed by Daston and Galison (2007). Furthermore, both consider the work of Hacking (1983) as crucial for understanding the difference between "representation" and "intervention." To the best of my knowledge this is perhaps the first attempt to extend the views of Daston and Galison (2007) in the area of research in chemistry, and to establish an explicit relationship between objectivity and transgression of categorization.

2. How exactly do you conceptualize Feyerabend's "epistemological anarchism" (Hoffmann, 2012, p. 36)?

   I see Feyerabend, whom I sadly never met, as a malevolent genius, intent on destruction of method. But underneath I sense in him an admiration for science, for the knowledge we gain, by hook, or by crook. I think of epistemological anarchism as the expression Feyerabend uses for the way science really works, namely

(a) that there is [are] no general rules governing scientific method,

(b) that there may be some protocols for gaining reliable knowledge (often learned by copying the attitudes of one's elders) in a given field, but that these are not necessarily recognized as foolproof or reliable by neighboring communities (even ones as close as chemical specialties—organic, physical)

A metaphor I have found useful recently is of scrabblers after knowledge. I mention it in the attached Tensions of Scientific Storytelling. (Hoffmann, 2016a)

My comments: Hoffmann (2014) develops the ideas of "compelling narratives" and "scrabblers after knowledge" as graphic representation of what scientists do, and then elaborates: "No anthropomorphization is needed. There is a life-giving tension between the several roles of the scientist as author, revealing and creating onion layers of reality's representation in his or her science" (p. 253).

3. In general, to what extent do your ideas on progress in science resonate with those of Daston and Galison (2007)?

Sadly, I have not yet read their book. Though I did read their essay in Representations predating the book by some years. I share their description of the evolution of attitudes toward images, but I wonder if they miss the special sense of realism that comes to chemists through their multisensual (sight, touch, smell) manipulation of compounds in the laboratory.

I think of science as the gaining of reliable knowledge. That knowledge is always contingent, dependent on the theories of the time. Yet through experimental manipulation (synthesis in chemistry is central) it gains substance and reality. (Hoffmann, 2016a)

My comments: "*Representations*" refers to Daston and Galison (1992) in which these authors presented the historical evolution of objectivity for the first time prior to their major publication in 2007. Hoffmann (2012, p. 28) has clarified that he prefers Ziman's (1978) "reliable knowledge" rather than van Frassen's concept of "empirically adequate."

4. It seems that although you recognize the importance of molecules such as C_{60} you are somewhat skeptical of nanochemistry? (p. 37).

I love these nano objects, but not the hype around them. And, to get back to Daston and Galison, and those images of molecules we have all seen from STM/AFM measurements—I think it's not at all clear what we are seeing. I send along another American Scientist column which muses on the meanings of images of the nanoworld. (Hoffmann, 2016a)

My comments: Hoffmann (2006) has referred to STM and AFM measurements in the following terms: "Are they faithful images? Not really. But neither are 'real' photographs, as anyone knows who has developed her own film or tinkered with an image electronically in a computer ..." (p. 2). Hoffmann's skepticism is justified as the images from STM/AFM are computer generated and not real photographs. Actually, in their enthusiasm, many general chemistry textbook authors and even perhaps some researchers do not clarify this difference.

6.2.3 Approaching a Crossroads

In order to facilitate a better understanding of the idea of being at cross-roads, I sent the following question to Professor Hoffmann on February 24, 2016: "Given the tension between representation and intervention, do you think that both science and science education, in a sense, are at the cross-roads?" Although, I do not entirely agree with him, in order to represent his viewpoint, I reproduce here the following response sent by Professor Hoffmann (Email to author, February 24, 2016b):

> I may not give the answer you want. I try to avoid or evade crises, and believe in the power of the movable middle. The tension between representation (passive) and intervention (active) has always been there in our central science. And in science education, though the active part there (laboratory instruction) has tended to be limited in the last decades, with growing safety concerns. Theory and representation has always been favored in teaching, and I think that is a problem for science education. I don't think we are at a crossroads. But I think we need more laboratory instruction, perhaps virtual reality will provide a poor substitute.

In my opinion, Professor Hoffmann's response in a certain sense is understandable (and I appreciate his frank statement), as he considers that the tension between representation (passive) and intervention (active) has always been a part of chemistry and that theory and representation have always been emphasized in teaching science. Nevertheless, in chemistry (also science) education, the historical evolution of objectivity (Daston & Galison, 2007) and the differentiation between representation and intervention in the context of nanotechnology, and the ensuing tension, are fairly novel concepts. Consequently, it is worthwhile to examine the role of interactions between objectivity, scientific method, nanotechnology, and the conflicts produced as we proceed from representation to presentation.

6.3 Progress in Science at a Crossroads

This juncture which has led us to a crossroads requires the ability of the scientists to interpret and understand, and Hoffmann (2012) has referred to it in cogent terms:

> Indeed, the interpretative skill of a scientist is one of the reasons why science—by contrast to what some historians and philosophers appear to believe—goes beyond, way beyond merely the following of a prescribed procedure, that would lead anyone well-versed in the "*scientific method*" from observations to conclusions. Leaps of the imagination do occur, and they are as important to the scientist as they are to the artist. (p. 200, italics added)

This criticism of the scientific method (i.e., prescribed procedure) is valid not only for historians and philosophers but also science educators and textbook authors. Historian of chemistry, Trevor Levere (2006) has referred to this problem in the following terms: "… many authors of science textbooks still write as if there were such a thing as *the* scientific method, and use labels like induction, empiricism, and falsification in simplistic ways that bear little relation to science as it is practiced" (pp. 115–116, original italics).

At this stage, it would be interesting to go back to Hoffman's (2012) endorsement of Feyeraband's (1975) "epistemological anarchism" referred to above. In a letter written to I. Lakatos, dated January 20, 1972, Feyerabend refers to the origin of his ideas with respect to "epistemological anarchism." He recounts that in 1965 while discussing the example of Brownian motion with a colleague it occurred to him that even such a highly confirmed theory could have an alternative. This would require the invention of new methods rather than adapting to "reason," and besides that he also acknowledged the role played by his Wittgensteinian upbringing. It is plausible to suggest that the idea of "transgression" was embedded in this episode. In the same letter, Feyerabend also referred to the scientific method in the following terms: "… the pleasant surprise I got when Sir Karl [Popper], then Prof. P., started his lectures on scientific method (in 1952) by saying: 'I am Professor of Scientific Method; but there is no scientific method …' which I liked …" (Reproduced in Motterlini, 1999, p. 272). The idea of transgression is much more clearer although somewhat implicit in *Against Method*: "… there is not a single rule, however plausible, and however firmly grounded in epistemology, that is not violated at some time or other …. Such violations are not accidental events, they are not results of insufficient knowledge or of inattention which might have been avoided. On the contrary, we see that they are necessary for progress" (Feyerabend, 1975, p. 23). In another letter written to I. Lakatos dated July 25, 1969, Feyerabend stated, "… the 'objectivity' of science is just moonshine?" (Reproduced in Motterlini, 1999, p. 169). Interestingly, Preston (1997) considers that there is a strong connection between the ideas of Feyeraband and the work of Michael Polanyi.

Science educators have also recognized the problematic nature of the scientific method. According to Grandy and Duschl (2007), the logical empiricist conception of science is closely related to the traditional scientific method, namely: make observations, formulate a hypothesis, deduce consequences from the hypothesis, make observations to test the consequences, and accept or reject the hypothesis based on the observations. In a study based on pre-service secondary science teachers' views related to scientific method, Windschitl (2004) concluded that these appear consistent with a "folk theory" of an atheoretical scientific method, "… that is promoted subtly, but pervasively, in textbooks, through the media, and by members of the science education community themselves" (p. 481). Niaz and Maza (2011) analyzed 75 general chemistry textbooks (published in USA) and found that only four textbooks facilitated satisfactorily an understanding that went beyond the traditional step-wise scientific method.

Niaz (2011, Chap. 9) has designed a study to facilitate in-service chemistry teachers' understanding of the scientific method and objectivity (among other aspects of the nature of science). A basic premise of the study was that a discussion of chemistry content within a historical context could help teachers to discuss, argue for or against a particular interpretation of experimental evidence, and finally deepen their understanding of various aspects of how science is practiced. For most teachers at the beginning of the course, scientific method provided a simple and straightforward way to understand how science is practiced. This conceptualization slowly started to change and most participants at the end of the course realized that this was a "caricature" of what real science is. The following are two examples of teachers' responses that were considered to be informed views of the scientific method (reproduced in Niaz, 2011, p. 142): "(a) In view of the universality and rigidity of the scientific method, one could believe that 'Science does not change'. For some it may signify that if science changes, *it does not exist*" (emphasis in the original) and "(b) Chemistry needs to be 'freed' of myths and history and philosophy of science could help. It needs to be emphasized that there is no one scientific method, but rather diverse methods and processes—textbooks cannot continue to be a list of questionnaires and algorithmic problems and answers."

Introducing objectivity of science in the classroom is a complex issue. In this study (Niaz, 2011, Chap. 9) in-service chemistry teachers were provided the following question/dilemma:

> Martin Perl, Nobel laureate in physics 1995, in his search for the fundamental particle (quark) has elaborated a philosophy of speculative experiments: "Choices in the design of speculative experiments usually cannot be made simply on the basis of pure reason. The experimenter usually has to base her or his decision partly on what feels right, partly on what technology they like, and partly on what aspects of the speculations they like" (Perl & Lee, 1997, p. 699). Given the methodologies of Thomson, Rutherford, Bohr, Millikan and Ehrenhaft (Niaz, 1998, 2000), in your opinion, what are the implications of this statement for teaching chemistry? (Reproduced in Niaz, 2011, p. 132)

The reference to Perl's experimental methodology is important as some students may think that what scientists did in the early twentieth century (e.g., Thomson, Rutherford, Bohr, and Millikan) was perhaps very different from what scientists do these days. Most of the teachers drew positive conclusions for teaching chemistry and following is one of the examples:

> Millikan did not manifest in public the speculative part of his research ... Perl, however has affirmed publicly that at times he speculates ... Perl's affirmation manifests what Millikan in some sense tried to "conceal," viz., *science does not develop by appealing to objectivity in an absolute sense* and that science does not have an explanation for everything and hence the need for research. Acceptance of the fact that science does not have an absolute truth and nor an immediate explanation for everything, would change students' conception of science and chemistry in particular. This will show chemistry to be a science in constant progress and that what is true today may be false tomorrow and may even help to originate a new truth—sequences of heuristic principles. (Reproduced in Niaz, 2011, p. 140, italics added)

This is an interesting response and some of its salient features are the following: (i) The question itself makes no mention of objectivity and still the teacher stated, "science does not develop by appealing to objectivity in an absolute sense." It seems plausible that the mention of lack of "pure reason" and "aspects of the speculations" led the teacher to think that the creative process needs to be flexible and hence the reference to "objectivity in an absolute sense"; (ii) Millikan's methodology (oil drop experiment, Holton, 1978a, b) is compared to that of Perl and that Millikan tried to conceal some of his data reduction procedures. Millikan-

Ehrenhaft controversy was one of the historical episodes discussed in class; (iii) The reference to "absolute truth" and that "what is true today may be false tomorrow" clearly shows the tentative nature of science; (iv) and that science cannot explain everything. This clearly shows that the context of chemistry content if presented within a history and philosophy of science perspective can facilitate students' and teachers' understanding of not only nature of science but also specific aspects such as the scientific method and objectivity (cf. Niaz, 2016).

These responses show that both the scientific method and objectivity can form part of classroom discussions more explicitly and thus provide a deeper understanding of these aspects. Consequently, Niaz (2016) designed a study based on in-service science teachers in which historical episodes were discussed and the following question was asked:

> Many students, professors, science textbooks and methodology courses emphasize the scientific method. Do you think the scientific method always plays a primordial role in scientific development? (Niaz, 2016, p. 59)

As compared to the previous study (Niaz, 2011), in this case the question explicitly refers to the role played by the scientific method. Out of the 12 teachers in the study, four responded that the scientific method is primordial, two responded that it is partially primordial and six responded that it is not primordial. Following is an example of a response that considered the scientific method not to be primordial:

> Scientific method is an implement used by every investigator to obtain information related to a problem in the natural and social sciences. However, in the history of the natural sciences it has been found that the scientific method as understood in the scientific community has not been followed in a strict and rigorous manner. One example is the discovery of charge of the electron (1.601×10^{-19}C), based on the "oil drop experiment." There is evidence that Millikan discarded data obtained in his experiment, which means that he did not follow or respect the scientific method rigorously and still his findings are to this day accepted by the scientific community. (Reproduced in Niaz, 2016, p. 61)

This teacher recognized the importance of the scientific method in solving problems. However, based on experience gained in classroom discussions (in which the oil drop experiment and the role played by Holton's, 1978a, b, study was discussed), she/he recognized that the scientific method was not followed rigorously in this case. Furthermore, it recognizes that despite such handling of the data, Millikan's study is still recognized by the scientific community. More recently, Holton (2014b) has elaborated on Millikan's discarding of data from some oil drops in the following terms: "So even if Millikan had included *all* drops and yet had come out with the same result, the error bar of Millikan's final result would not have been remarkably small, but large—the very thing Millikan did not like" (original italics). It is plausible to suggest that such an understanding by teachers based on historical context can help them to introduce these topics in the classroom meaningfully, and thus motivate students, and facilitate conceptual understanding.

Before presenting the evaluation of general chemistry textbooks (next section), it is interesting to consider the following presentation of the oil drop experiment in a physics textbook (Olenick, Apostol, & Goodstein, 1985) that formed part of a study conducted within a history and philosophy of science perspective by Rodríguez and Niaz (2004a). This textbook went to considerable length (about 5 pages) to present Millikan's research methodology and pointed out the dilemmas and contradictions in the handling of the data: "By observing the motion of the hundreds of droplets with different charges on them, Millikan *uncovered the pattern he expected*: the charges were multiples of the smallest charge he measured" (Olenick et al., 1985, p. 241, italics added). "The pattern" can be considered as an oblique reference to Millikan's guiding assumption. The textbook reproduced the following quote from Millikan's laboratory notebook (dated March 15, 1912; see Holton, 1978a for Millikan's lab notebooks): "One of the best ever [data] ... almost exactly *right*. Beauty—publish" (original italics). After reproducing the quote, the textbook authors asked a very thought-provoking question: "What's going on here? How can it be right if he's supposed to be measuring something he doesn't *know*? One might expect him to publish everything!" (Olenick et al., 1985, p. 244, original italics). These are important issues related to understanding science, namely can a scientist know beforehand what he is going to find, and what is even more difficult to understand is that how can a scientist know the right answer before doing the experiment. Interestingly, the authors themselves provided further insight and advice for students:

> Now, you shouldn't conclude that Robert Millikan was a bad scientist .... What we see instead is something about how real science [cutting-edge] is done in the real world. What Millikan was doing was not cheating. He was applying scientific judgment .... But experiments must be done in that way. Without that kind of judgment, the journals would be full of mistakes, and we'd never get anywhere. So, then, what protects us from being misled by somebody whose "judgment" leads to wrong results? Mainly, it's the fact that someone else with a different prejudice can make another measurement .... Dispassionate, unbiased observation is supposed to be the hallmark of the scientific method. Don't believe everything you read. Science is a difficult and subtle business, and there is no method that assures success. (Olenick et al., 1985, p. 244)

If we compare this presentation with the traditional flow diagram of the scientific method found in most textbooks, the role played by historical reconstructions constitutes an important source of understanding how science is practiced. This is a good illustration of how Millikan's presuppositions and heuristic principle can facilitate students' understanding of "science in the making" and eventually nature of science, based on: (a) Doing experiments means gathering data and its interpretation (scientific judgment); (b) Without such judgments journals would be full of mistakes; (c) Some scientists can be misled in their judgments; (d) Another scientist with a different heuristic principle can present an alternative interpretation, namely science is self-correcting; (e) There are no dispassionate, unbiased observations as suggested by the scientific method, namely observations are theory-laden. At this stage it is important to consider if after following this methodology, Millikan was being "objective" in the handling of his data. A possible response can be found in what Hoffmann (2012) has suggested, namely science, "... goes beyond, way beyond merely the following of a prescribed procedure" (p. 200). This also illustrates what Matthews (1992) has referred to as "historian's theory."

In this quest for understanding what scientists do, Hoffmann (2014) has presented a graphic representation of the roles played by the scientist that include "compelling narratives" and "scrabblers after knowledge" in the following terms:

> Carefully done measurements of observables are an essential ingredient of science, against which theories must be measured. They constitute facts, some will say. Well, *facts are mute. One needs to situate the facts, or interpret them.* To weave them into nothing else but a narrative. The *tension of the scientific narrative* resides in the divided personality (or personalities) of the authors, scrabbler and writer, and the representation of reality that their work shapes. *Reality turns a different crystal face to all its viewers.* With the writer telling the neat story that the stumbling yet imaginative scrabbler found, the investigators together build reality, or a face of reality. That face is in turn seen in a different light by others who compete with, or who follow, the one person who is both scrabbler and writer. (p. 252, italics added)

Indeed, this represents the scientific endeavor in succinct terms in which the scientist goes through the phases of a scrabbler and a writer to convince others of his "narrative" which involves an understanding of the following: facts are mute and hence need interpretation, reality presents a different perspective to different scientists and hence progress in science is based on narratives that generate tensions.

6.4   Criteria for Evaluation of General Chemistry Textbooks

In the previous sections of this chapter, I have shown that objectivity needs to be understood within a historical context and that the scientific method does not necessarily play an important role in scientific progress. Similarly, the development of nanotechnology (STM, AFM, others) has led to a deepening of our understanding of how scientific progress is at a crossroads due to the differentiation between representation and presentation (intervention). Furthermore, it is plausible to suggest that there could be a relationship between objectivity, scientific method, and the juncture that has led us to a crossroads. In other words, it would be interesting to observe if those textbooks that recognize the problematic nature of objectivity also accept that there is no universal step-wise scientific method. All general chemistry textbooks analyzed in this chapter were published in USA between 1990 and 2017 (Appendix 9 provides a complete list and references of all the textbooks). With this background following criteria were developed for evaluating general chemistry textbooks:

6.4.1   Criterion 1: Objectivity

Following classifications were used for evaluating textbooks:

Satisfactory (S): Textbook explicitly recognizes the problematic nature of objectivity and that as an epistemic virtue it is achieved in degrees based on controversies and interactions among members of the scientific community.

Mention (M): Simple mention with no details of the underlying issues.

No mention (N): Problematic nature of objectivity and the underlying issues are ignored.

6.4.2   Criterion 2: Scientific Method

Following classifications were used to evaluate textbooks:

Satisfactory (S): Progress in science depends on imagination, creativity, controversies, and not on a prescribed set of steps. Furthermore, success in science is not achieved by simply following a series of procedures similar to a recipe book, but rather understanding the different facets of reality.

Mention (M): Following strict rules of procedure does not automatically lead to progress in science. One example would be the following: scientific method is not a rigid sequence of steps, but rather a dynamic process designed to explain and predict real phenomena.

No mention (N): No mention of the issues involved, although the textbook may still refer to scientific method.

6.4.3   Criterion 3: Scanning Tunneling Microscopy (STM)

Following classifications were used to evaluate textbooks:

Satisfactory (S): Scanning tunneling microscopy (STM) helps in "seeing," creating, and manipulating to construct a new world of atom-sized products. However, it does not provide photographs but rather computer-generated images. This distinction is important if we want to understand the difference between representation and presentation (intervention).

Mention (M): STM helps to "see" (representation) and manipulate individual atoms to provide magnifications (presentation). However, it does not differentiate if STM provides photographs or computer generated images.

No mention (N): No mention of STM or related issues.

6.4.4   Criterion 4: Atomic Force Microscopy (AFM)

Following classifications were used to evaluate textbooks:

Satisfactory (S): Atomic force microscopy (AFM) helps in seeing, creating, and manipulating to construct a new world of atom-sized products. However, it does not provide photographs but rather computer-generated images. This distinction is important if we want to understand the difference between representation and presentation (intervention).

Mention (M): AFM helps to "see" (representation) and manipulate individual atoms to provide magnifications (presentation). However, it does not differentiate if AFM provides photographs or computer generated images.

No mention (N): No mention of AFM or related issues.

6.4.5   Criterion 5: From Representation to Presentation: Scientific Progress at a Crossroads

Following classifications were used to classify textbooks:

Satisfactory (S): Study and depict what actually exists (representation) and also indicate what can be manipulated (presentation) to produce new products as part of nanotechnology.

Mention (M): A simple mention of some applications of STM and AFM, with no reference to new products and their manipulation.

No mention (N): No mention of STM, AFM, and related issues.

Following the guidelines based on Charmaz (2005), presented in Chap. 3, and in order to facilitate credibility, transferability, dependability, and confirmability (cf. Denzin & Lincoln, 2005) of the results, I adopted the following procedure: (a) All the 60 general chemistry textbooks were evaluated and classified in one of the three levels: satisfactory, mention, and no mention (as compared to studies in Chaps. 3–5, these were not assigned to categories); and (b) After a period of approximately 3 months all the textbooks were evaluated again and there was an agreement between the first and the second evaluation: 91% on Criterion 1, 89% on Criterion 2, 94% on Criterion 3, 92% on Criterion 4, and 95% on Criterion 5.

6.5   Results and Discussion

In this section, I report results of the evaluation of general chemistry textbooks based on the five criteria presented in the previous section. Table 6.1 shows that understanding objectivity (Criterion 1) was the most difficult for general chemistry textbooks as 90% were classified as no mention (N), and only 8% as satisfactory (S). It is interesting to observe that as compared to objectivity (8%), textbooks understood better the importance of STM (Criterion 3, Satisfactory 27%), and scientific progress at a crossroads (Criterion 5, Satisfactory 25%). This shows the difficulties involved in understanding objectivity.

However, it is important to note (see Appendix 10) that those textbooks that were classified as satisfactory (S) on Criterion 1 (objectivity) were also classified as satisfactory on Criterion 2 (scientific method). Apparently, understanding objectivity also leads to a better understanding of scientific method. Appendix 10 also shows that only 10 textbooks (out of the 60 included in this chapter), had a score of 50% (6 or more points). Now I present examples of textbook presentations related to the five criteria.

Table 6.1 Distribution of general chemistry textbooks according to criteria and classification ($n = 60$)

Criteria	Classification		
	N (%)	M (%)	S (%)
1	54 (90)	1 (2)	5 (8)
2	29 (48)	20 (33)	11 (18)
3	28 (47)	16 (27)	16 (27)
4	48 (80)	5 (8)	7 (12)
5	34 (57)	11 (18)	15 (25)

Notes: N = No mention, M = Mention, S = Satisfactory. Criterion 1 = Objectivity, Criterion 2 = Scientific method, Criterion 3 = STM, Scanning tunneling microscopy, Criterion 4 = AFM, Atomic force microscopy, and Criterion 5 = Scientific progress at a crossroads.

6.5.1 Criterion 1: Objectivity

One textbook stated the following with respect to data collection in science:

> The data normally must be collected under conditions that can be reproduced anywhere in the world. Then new data can be obtained to confirm or to refute the correctness of the suggested pattern. The results represent a unique type of *objective truth* that is ideally independent of differences in the language, culture, religion, or economic status of the various observers. Such established truth is appropriately referred to as **scientific fact** (Joesten, Johnston, Netterville, & Wood, 1991, p. 6, emphasis in the original, italics added). Classified as M.

Perhaps most science teachers and textbook authors would agree with this presentation that was classified as Mention (M). Nevertheless, history of science shows that science in the making (i.e., establishment of a scientific fact) is extremely complex and very different from what it is generally believed to be (cf. Niaz, 2012). Furthermore, what is considered as a scientific fact today may not be considered to be so by later scientists. Of course, it can be argued that such issues need not be discussed in introductory chemistry (science) courses. Precisely, recent research in science education has emphasized the importance of such controversial aspects as part of nature of science and that students be made aware of it (cf. Hodson, 2009).

Following are examples of textbooks that were classified as Satisfactory (S):

> Kuhn's ideas created a controversy among scientists and science historians that continues to this day. Some, especially postmodern philosophers of science, have taken Kuhn's ideas one step further. They argue that scientific knowledge is *completely* biased and lacks any objectivity. Most scientists, including Kuhn, would disagree .... In other words, saying that science contains arbitrary elements is quite different from saying that science itself is arbitrary. (Tro, 2008, p. 7, original italics)

> However, it is important to remember that science does not always progress smoothly and efficiently. Scientists are human. They have prejudices; they misinterpret data; *they become emotionally attached to their theories and thus lose objectivity*; and they play politics. Science is affected by profit motives, budgets, fads, wars, and religious beliefs. (Zumdahl & Zumdahl, 2014, p. 8, italics added)

You many think that research in science is straightforward: Do experiments, collect information, and draw a conclusion. But, research is seldom that easy. Frustrations and disap-
pointments are common enough, and *results can be inconclusive*. Experiments often
contain some level of uncertainty, and *spurious or contradictory data* can be collected.
For example, suppose you do an experiment expecting to find a direct relation between
two experimental quantities. You collect six data sets. When plotted on a graph, four of
the sets lie on a straight line, but two others lie far away from the line. Should you ignore
the last two data sets? Or should you do more experiments when you know the time they
take will mean someone else could publish results first and then get the credit for a new
scientific principle? Or should you consider that the two points not on the line might indicate that *your original hypothesis is wrong* and that you will have to abandon a favorite
idea you have worked on for many months? Scientists have responsibility *to remain
objective in these situations, but sometimes it is hard to do*. (Kotz, Treichel, Townsend, &
Treichel, 2015, p. 5, italics added). Classified as Satisfactory (S)

The presentation by Kotz et al. (2015) succinctly presents the dilemmas
involved in doing scientific research. Most scientists have confronted with situations in which *results can be inconclusive*, data that are *spurious or contradictory*,
original hypothesis is wrong and despite the heavy odds it is still important to be
objective. History of science provides ample evidence of many episodes in which
scientists had to deal with situations in which it was difficult to be objective
(cf. Niaz, 2009, 2012). In his determination of the elementary, electrical charge
Robert Millikan had to face a similar dilemma and he wrote: "*It will be seen from
Figs. 2 and 3 that there is but one drop in the 58 whose departure from the line
amounts to as much as 0.5%. It is to be remarked, too, that this is not a selected
group of drops but represents all of the drops experimented upon during 60 conse-
cutive days*" (Millikan, 1913, p. 138, original italics). As suggested by Kotz et al.
(2015), scientists are concerned with respect to all data points lying on a line.
Millikan was happy as only one drop of the 58 deviated from the line. However,
the problematic nature of this statement became clear many years later when Holton
(1978a, b) studied Millikan's hand-written notebooks and found that he did not
experiment with 58 drops but 140. In other words, Millikan discarded data from 82
(59%) of the drops. Was Millikan being objective and if so how do we explain this
to our students? To facilitate an understanding of this dilemma, Niaz and Rivas
(2016) have designed a classroom teaching strategy for high school students.

6.5.2 Criterion 2: Scientific Method

An important guideline for understanding the scientific method is the following
approach suggested by Hoffmann (2014): "By analyzing exactly how scientists
approach scientific literature, I hope to reveal the humanity of the scientific
method" (p. 323).

Following is an example of a textbook that was classified as No mention (N):

Often experiments are designed to try to confirm or support a hypothesis or to disprove it.
One valid experiment that disproves a hypothesis may be enough to reject it, or at least
require its alteration. Thomas Henry Huxley, an English biologist, in 1870 gave eloquent

testimony to this idea when he said, "The great tragedy of Science [is] the slaying of a beautiful hypothesis by an ugly fact." (Dickson, 2000, p. 4, italics added)

Interestingly, despite Huxley's merits as a biologist his views on philosophy of science (role of crucial experiments) are not in consonance with the history of science or modern philosophers of science. Early in the twentieth century, Duhem (1914) emphasized that an experiment would be "crucial" only if it conclusively eliminated every possible set of hypotheses (also see Losee, 2001). A good example to illustrate this point is the Michelson-Morley experiment. When first performed in 1887 it provided a "null" result with respect to the ether-drift hypothesis, viz., no observable velocity of the earth with respect to the ether. Following Huxley, this experimental evidence should have been crucial in rejecting the ether-drift hypothesis. However, this was not the case. Michelson (a Nobel laureate in physics) and colleagues continued to perform further experiments to prove the hypothesis and even organized an international conference as late as 1927 to evaluate the latest experimental evidence (for details see Niaz, 2009, Chap. 2). According to Lakatos (1970, p. 162), it took the scientific community almost 25 years to consider the Michelson-Morley experiment as the greatest negative experiment in the history of science.

Let us now compare the above presentation with the following that was classified as Satisfactory (S):

The law of multiple proportions was not known before Dalton presented his theory, and its discovery demonstrates the scientific method in action. Experimental data suggested to Dalton the existence of atoms, and the atomic theory suggested the relationship that we now call the law of multiple proportions. Repeated experimental tests have uncovered no instances where the law of multiple proportions fails. These successful tests added great support to the atomic theory. In fact, for many years the law was one of the strongest arguments in favor of the existence of atoms (Brady & Senese, 2009, p. 39, previously on page 3 authors had stated: "Scientific method helps us build models of nature."

It is important to note that instead of the flow diagram (presented in most textbooks) these authors have used the context of a historical episode to illustrate how the scientific method works. The role played by Dalton's atomic theory and its relationship to the law of multiple proportions has been the subject of considerable controversy among historians and philosophers of science. For example, both Needham (2004) and Chalmers (2009) have argued that the progress made by chemistry in the nineteenth century owed little to Daltonian atomism. In contrast, Rocke (2013) has claimed just the opposite: "I propose to rescue nineteenth-century atomic theory from the charge of irrelevance or even meaninglessness. I claim that atomic theory was, from the beginning, not only a robust and heuristically powerful theory but also crucial to the spectacular development of chemistry in that century" (p. 146). In this context, the presentation by Brady and Senese (2009) is even more significant and also provides textbook authors an opportunity to introduce the scientific method in a more meaningful manner. Interestingly, in a recent study, Niaz (2016, Chap. 4) found that of the 32 general chemistry textbooks (published in USA) analyzed, 19 either ignored the issues involved or reiterated that Dalton was led to his atomic theory by the discovery of the law of multiple proportions (a historical reconstruction shows just the opposite). This provides a good example of what Matthews (1992) has referred to as historians' theory.

The textbook by Ebbing (1996) illustrated the different steps of the scientific method by comparing them with Rosenberg's discovery of the anticancer activity of cisplatin. A similar presentation is included in Ebbing and Gammon (2013, 2017). All three presentations were classified as Mention (M), and could have been classified as Satisfactory (S) if more details of Rosenberg's work had been included especially the creative part. As presented in these textbooks, it seems that Rosenberg was more concerned about the next step of the scientific method rather than his research program.

Following is an example of a textbook that was classified as Mention (M):

> The sequence of steps just described—constitute the **scientific method**—From the preceding discussion you may get the impression that scientific progress always proceeds in a dull, orderly, and stepwise fashion. This isn't true; science is exciting and provides a rewarding outlet for cleverness and creativity. Luck, too, sometimes plays an important role. For example, in 1828 Frederick Wöhler, a German chemist, was heating a substance called ammonium cyanate in an attempt to add support to one of his hypotheses. His experiment, however, produced an unexpected substance, which out of curiosity he analyzed and found to be urea (a constituent of urine). This was an exciting discovery, because it was the first time anyone had knowingly ever made a substance produced only by living creatures from a chemical not having a life origin. The fact that this could be done led to the beginning of a whole branch of chemistry called *organic chemistry*. Yet, had it not been for Wöhler's curiosity and his application of the scientific method to his unexpected results, the significance of his experiment might have gone unnoticed. (Brady, Russell, & Holum, 2000, p. 3, emphasis and italics in the original)

This is an interesting presentation that first emphasizes the importance of the scientific method and then goes on to illustrate the role played by creativity by drawing on a historical episode. However, this was not classified as Satisfactory (S) as it associates Wöhler's discovery with the scientific method, which is precisely what was not of much help in this particular case.

Silberberg (2000) summarized the scientific method as: "The scientific method is not a rigid sequence of steps, but rather a dynamic process designed to explain and predict real phenomena" (p. 13). However, earlier in the chapter this textbook stated: "An experiment is a clear set of procedural steps that tests a hypothesis. Experimentation is the connection between our ideas, or hypotheses, about nature and nature itself. Often, hypothesis leads to experiment, which leads to revised hypothesis, and so forth. *Hypotheses can be altered, but the results of an experiment cannot*" (p. 12, italics added). Except for the part in italics this is a fairly good representation of the scientific method. History of science shows that what counts as results of an experiment can vary from one scientist to another. Determination of the elementary electrical charge showed that what counted as results varied in the experimental work of Robert Millikan and Felix Ehrenhaft (for details see Holton, 1978a, b; Niaz, 2009). Actually, in general "science in the making" is a better indicator of what happens in the laboratory than what even the scientists themselves report after the experiment has concluded (cf. Niaz, 2012). This textbook was classified as Mention (M).

Malone and Dolter (2013) provided the following historical episode in order to illustrate various aspects of the scientific method:

In 1979, American scientists discovered a thin layer of sediment in various locations around the world that was deposited about 65 million years ago, coincidentally the same timeframe in which the dinosaurs became extinct. Indeed, there were dinosaur fossils below that layer but none above. Interestingly, that layer contained comparatively high amounts of iridium. Scientists proposed that this layer contained the dust and debris from a collision of a huge asteroid or comet (about 6 miles in diameter) with Earth. They concluded that a large cloud of dust must have formed, encircling Earth and completely shutting out the sunlight. A bitter cold wave followed, and most animals and plants quickly died. A hypothesis (a tentative explanation of facts) was proposed that the dinosaurs must have been among the casualties (p. 57, in a section entitled "Iridium, the missing dinosaurs and the scientific method"). This presentation was classified as Mention (M).

This presentation goes beyond the traditional pattern found in most textbooks by illustrating the various aspects of the scientific enterprise in the context of a historical episode. Some of the salient features of this presentation are: (a) Iridium layer was found at various locations around the world; (b) Formation of the layer coincided with the extinction of the dinosaurs; (c) Dinosaur fossils were always found below the layer; (d) Postulation of an hypothesis to explain a collision between an asteroid and the earth leading to the extinction of the dinosaurs. Of course, inclusion of the role played by the scientific community in accepting this hypothesis and its skepticism could have provided a better picture of the scientific progress.

Olmsted and Williams (2006) did not include the traditional flow diagram to present the scientific method. Instead, these authors included a diagram starting in 1971 and related events that led by 1987 to the awareness about CFC (chlorofluorocarbons) as a possible cause of the depletion of atmospheric ozone. This is a fairly good illustration of how scientific understanding goes through the different stages of development, and was classified as Mention (M).

Brown, LeMay, Bursten, Murphy, and Woodward (2014) present the scientific method as a traditional sequence of steps based on a flow diagram. However, in the same section they included the following statement:

As we proceed this text, we will rarely have the opportunity to discuss the *doubts, conflicts, clashes of personalities, and revolutions of perception* that have led to our present scientific ideas. You need to be aware that just because we can spell out the results of science so concisely and neatly in textbooks does not mean scientific progress is smooth, certain, and predictable. (p. 27, italics added)

This textbook espouses the traditional scientific method and still emphasizes the importance of "doubts, conflicts, clashes of personalities and revolutions of perception," in scientific progress. Indeed, this is an innovative step toward depicting how "science in the making" goes beyond the traditional scientific method and textbooks can play an important role in bringing this to our students' attention. Furthermore, textbooks do not have to go into detailed descriptions of historical episodes, but rather present some concise accounts of what has been well established. Brown, LeMay, Bursten, and Murphy (2009) also presented a similar statement and both textbooks were classified as Mention (M).

Following is an example of a textbook that emphasized the ability to ask questions by referring to the life and work of Linus Pauling in the following terms:

When the late Nobel Laureate Linus Pauling described his student life in Oregon, he recalled that he read many books on chemistry, mineralogy, and physics. "I mulled over the properties of materials: why are some substances colored and others not, why are some minerals or inorganic compounds hard and others soft?" He said, "I was building up this tremendous background of empirical knowledge and at the same time asking a great number of questions." Linus Pauling won two Nobel Prizes: the first, in 1954, was in chemistry for his work on the structure of proteins; the second in 1962, was the Peace Prize. (Timberlake, 2010, p. 4)

Furthermore, this textbook (classified as Mention, M) besides presenting the flow diagram (p. 5) also presented the following statements (p. 5) for the students to classify as an observation, hypothesis, or an experiment: (a) Drinking coffee at night keeps me awake (observation); (b) When I drink coffee only in the morning, I can sleep at night (observation); (c) I will try drinking coffee only in the morning (experiment); (d) A silver tray turns a dull grey when left uncovered (observation); (e) It is warmer in summer than in winter in the northern hemisphere (observation); (f) Ice cubes float in water because they are less dense (hypothesis). These questions help students to understand the different aspects of the scientific method (as presented in flow diagrams) in a more meaningful and realistic context. Actually, research has shown that students and even science teachers have considerable difficulty in classifying such statements. For example, Cortéz and Niaz (1999) studied adolescents' (6th to 11th grade, 11–17 years old) understanding (classifying statements) of *observation, prediction, and hypothesis* and found that 11th grade students (with the best performance) obtained a mean score of 47.6% on everyday items and of 37.3% on educational items. In a subsequent study, (Niaz, 2011, Chap. 7) almost 50% of the teachers had considerable difficulty in classifying hypotheses and predictions and some teachers explicitly elaborated and classified a prediction as a hypothesis.

Blei and Odian (2006) presented the following example to illustrate how the scientific method worked in a real situation (accompanied by the traditional flow diagram):

In 1928, it was discovered that a nonpathogenic strain of pneumococcus could be transformed into a virulent strain by exposure to chemical extracts of the virulent strain. Call this discovery a fact or an observation. The bacteria is *Diplococcus pneumoniae*, and the virulent strain causes pneumonia .... The material in these extracts responsible for the transmittance of inheritance was called "transforming principle," but its chemical nature was unknown. To uncover the chemical identity of the transforming principle, scientists required a hypothesis, a guess or hunch .... Most biochemists at that time believed that inheritance was carried by proteins, and that became the first hypothesis proposed .... [Experiments helped to discard this hypothesis] .... An alternative testable hypothesis was proposed—that the transforming substance could be DNA .... [Experiments provided support for this hypothesis] .... Because of all the subsequent experimental support of the idea that DNA is the molecule that carries genetic information, it now has the status of a theory, a hypothesis in which scientists have a high degree of confidence (p. 4, italics in the original). Classified as Mention (M).

This is an interesting presentation of an historical episode in which different aspects of the dynamics of scientific progress are manifested, namely role of alternative hypotheses, experimental evidence to support a hypothesis, and prior beliefs of the scientists (inheritance carried by proteins). Contrary to what the authors suggest this precisely shows that scientists have to be imaginative and creative, which means going beyond the scientific method.

Following is an example of a textbook that was classified as Satisfactory (S):

> One last word about the scientific method: some people wrongly imagine science to be a strict set of rules and procedures that automatically lead to inarguable, objective facts. This is not the case. Even our diagram of the scientific method is only an idealization of real science, useful to help us see the key distinctions of science. Doing real science requires hard work, care, creativity, and even a bit of luck. Scientific theories do not just fall out of data—they are crafted by men and women of great genius and creativity. A great theory is not unlike a master painting and many see a familiar kind of beauty in both. (Tro, 2008, p. 6)

This is a fairly good presentation of the difficulties involved in doing research and hence provides teachers an opportunity to go beyond the traditional recipe-like scientific method. Of course, this could have been accompanied with some episodes from the history of science.

The textbook by Denniston, Topping, and Caret (2011) presented the traditional scientific method as a flow diagram (p. 4). However, at the same time these authors included a description of how Alexander Fleming discovered penicillin. Next, in a section entitled, "Curiosity, Science, and Medicine" stated: "Curiosity is also the basis of the scientific method" (p. 3), then related the experience of Michael Zasloff in the discovery of *magainins* found in the skin of frogs. While working at the National Institute of Health (USA) Zasloff's experiments involved the surgical removal of the ovaries of African clawed frogs. After surgery he put the frogs back in their tanks which were full of bacteria. However, to his surprise the frogs healed quickly and there was no infection. Of all the scientists working on the subject only Zasloff was curious enough to "speculate" that there could be chemicals in the frogs' skin that defended them against infection. Eventually, research showed that there were two molecules in the frog skin that killed the bacteria. Zasloff named them *magainins*, after the Hebrew word for shield. Such episodes from the history of science (medicine in this case) can provide students stimulating experiences and food for thought. This presentation was classified as Satisfactory (S).

In order to introduce the scientific method, one textbook (Bettelheim, Brown, Campbell, & Farrell, 2010) did not include a flow diagram, as presented in most textbooks. Instead it used the history of science to illustrate various facets of how science is done. For example, in order to show that science does not always follow the same path, such as: facts first, hypothesis second, theory last, it included Mendeleev's prediction of the element germanium in 1871, before it was actually discovered in 1886 (for details with respect to Mendeleev's contribution in the elaboration of the periodic table see Brito, Rodríguez, & Niaz, 2005). At the same time, the textbook clarified that in the history of science many firmly established theories were eventually discarded because they could not pass new tests (p. 4). Furthermore, this textbook emphasized the role of serendipity in the following terms:

> On the other hand, many scientific discoveries result from **serendipity**, or chance observation. An example of serendipity occurred in 1926, when James Sumner of Cornell University left an enzyme preparation of jack bean urease in a refrigerator over the weekend. Upon his return, he found that his solution contained crystals that turned out to be a protein. This chance discovery led *to the hypothesis that all enzymes are proteins*. Of course, serendipity is not enough to move science forward. Scientists must have the creativity and insight to recognize the significance of their observations. Sumner fought for more than 15 years for his hypothesis to gain acceptance because people believed that only small molecules can form crystals. Eventually his view won out, and he was awarded a Nobel Prize in chemistry in 1946 (Bettelheim, Brown, Campbell, & Farrell, 2010, emphasis in the original and italics added). Classified as Satisfactory (S).

This episode from the history of science is perhaps more illustrative of how a hypothesis is formulated in real science (*all enzymes are proteins*), than a flow diagram, and can help to arouse students' curiosity.

Another textbook innovated by presenting the scientific method in the context of a historical episode related to Lavoisier and phlogiston:

> We have selected a few of Antoine Lavoisier's early experiments to illustrate what has become known as the scientific method (Fig 1–5). Examining the history of physical and biological sciences reveals features that occur repeatedly. They show how science works, develops, and progresses. They include the following: ... [among others] Being skeptical. Lavoisier was skeptical of the phlogiston hypothesis because metals gained weight when strongly heated. If this process was similar to burning wood, why was phlogiston not lost? (p. 4). [Later the authors continue to present another aspect of how science works] .... Communication is not usually included in the scientific method, but it should be. Lavoisier knew about oxygen because he read the published reports of Joseph Priestley and Carl Wilhelm Scheele, who discovered oxygen independently in the early 1770s (Cracolice & Peters, 2016, pp. 4–5). Classified as Satisfactory (S).

The importance of Lavoisier-Priestley debate has also been discussed extensively in the science education literature (De Berg, 2011; Song & Young, 2014). Interestingly, this textbook includes a modified form of the flow diagram in which "skepticism" is included in the same box along with "predicting, testing and revising." Such presentations based on historical episodes can provide students with meaningful experiences and explore others. Many general chemistry textbooks discuss the Lavoisier-Priestley debate and also other historical episodes. However, most of such presentations do not highlight important facets of how science works, but rather present a rigid and cyclic method based primarily on observations, hypotheses, and experiments.

Brown and Holme (2011) adopted a critical approach in trying to explain scientific advancement: "The word 'method' implies a more structured approach than actually exists in most scientific advancement. Many of the advances of science happen coincidentally, as products of serendipity. The stops and starts that are characteristic of scientific development, however, are guided by the process of hypothesis formation and observation of nature. Skepticism is a key component of this process. Explanations are accepted only after they have been held up to the scrutiny of experimental observation" (p. 12). Interestingly, these authors did not include a flow diagram of the scientific method as found in most textbooks. Authors found the word "method" itself as problematic and this provides one alternative to the flow diagrams. The idea of "stops and starts" and skepticism is particularly helpful in understanding how formation of hypotheses and experimental observations are intricately linked and do not constitute a simple hierarchical structure. This presentation was classified as Satisfactory (S).

Various textbooks (Brady, Russell, & Holum, 2000; Brown et al., 2014; Ebbing & Gammon, 2013; Malone & Dolter, 2013) provide a fairly good reconstruction of historical episodes that illustrate the role of controversies, conflicts, clashes of personalities, alternative conceptions, and prior beliefs of scientists. However, after providing such examples these textbooks end up endorsing the traditional step-wise scientific method. One textbook (Brown & Holme, 2011) provided an alternative by pointing out that the word "method" is itself problematic as it denotes a more structured approach than actually found in the history of science. On the other hand, difficulties involved in the scientific endeavor (leading to "stops" and "starts") accompanied by skepticism are better ways to understand science. Some of the ideas presented in "How science works" (Undsci.berkeley. edu) can be helpful in understanding a more realistic picture of the scientific enterprise. Similarly, Binns and Bell (2015) have suggested that the reference to the scientific method itself be avoided and that instead teachers could emphasize the work of scientists within a historical context.

Hoffmann (2014) has emphasized that analyzing how scientists approach scientific literature reveals the humanity of the scientific method. The narrators (scientists) in chemical articles are human beings, although they may try to efface themselves by writing in the third person.

6.5.3 Criterion 3: Scanning Tunneling Microscopy (STM)

One of the textbooks in a section entitled "Critical Thinking" included the following question for the students to consider: "The scanning tunneling microscope allows us to 'see' atoms. What if you were sent back in time before the invention of the scanning tunneling microscope? What evidence could you give to support the theory that all matter is made of atoms and molecules" (Zumdahl & Zumdahl, 2014, p. 4, Classified as Mention, M). Such questions can arouse students' curiosity while trying to understand not only how the STM works but also its implications. Interestingly, a Nobel Laureate in chemistry has referred to the same question in the following terms: "… but we did not wait for scanning tunneling microscopes to show us molecules; we gleaned their presence, their stoichiometry, the connectivity of the atoms in them, and eventually their metrics, shape, and dynamics, by indirect experiments" (Hoffmann, 2012, p. 28). Discussion of such questions can facilitate students' understanding of how science progresses and at times scientists do not have all the means to resolve a problem, and still the quest for knowledge and progress continues. The textbook by Ellis, Geselbracht, Johnson, Lisensky, and Robinson (1993) included a reference to STM manufactured for classroom use (Burleigh Instruments, New York).

Following is an example of a textbook that was classified as Mention (M): "In other words, Binnig, and Rohrer had discovered a type of microscope that could 'see' atoms. Later work by other scientists showed that the STM could also be used to pick up and move individual atoms or molecules, allowing structures and patterns to be made one atom at a time" (Tro, 2008, p. 46). To illustrate this aspect of nanotechnology, Tro shows the Kanji characters for the word "atom" written with individual iron atoms on top of a copper surface, and also an STM image of iodine atoms on a platinum surface (p. 46). However, the author does not clarify if these are photographs or computer-generated images. Interestingly, Tro (2008) also states that after 200 years Dalton's atomic theory has been validated by the imaging of atoms by means of the STM (p. 6). Another example of a textbook that was classified as Mention (M) is provided by Russo and Silver (2002): "Atoms are so tiny that, until recently, scientists thought we would never be able to see them. They spoke too soon. Though no one has seen an atom through an ordinary microscope, in the early 1980s a device called scanning tunneling microscope produced the first images of individual atoms—like the silicon atoms that appear as bumps on the surface of the silicon crystal shown in the photograph at right" (p. 7). Following is another example of a textbook that was classified as Mention (M): "The *ability to manipulate individual atoms* has the potential to allow scientists to control reactions of single atoms and molecules. This could lead to the production of new chemical substances that are not possible using normal chemical methods" (Spencer, Bodner, & Rickard, 1999, p. 8, italics added).

Silberberg (2000) described STM as: "In practice, the tunneling electrons create a current that can be used to image the atoms of an adjacent surface. An extremely sharp tungsten-tipped probe, the source of the tunneling electrons, is placed very close (about 0.5 nm) to the surface under study. A small potential is applied across this minute gap to increase the probability that the electrons will tunnel across it. The size of the gap is kept constant by maintaining a constant tunneling current generated by the moving electrons. For this to occur, the probe must move tiny distances up and down, thus following the atomic contour of the surface. This movement is electronically monitored, and after many scans, a three-dimensional map of the surface is obtained" (p. 454). This presentation was classified as Mention (M).

Joesten, Johnston, Netterville, and Wood (1991) presented STM in the following terms: "The STM is an astonishing device because of its inherent simplicity .... By adjusting the up-down position of the tungsten needle as it moves across the surface, a constant tunneling current is maintained. As this takes place, however, the positions of the atoms are actually measured giving a picture of the atomic landscape" (p. 82). Such presentations leave the impression that STM provides actual photographs of the atoms, whereas the images are computer-generated. This presentation was classified as Mention (M).

In a section entitled, "A four-wheel-drive nanocar," Zumdahl and DeCoste (2015) stated:

> A special kind of "microscope" called a scanning tunneling microscope (STM) has been developed that allows scientists to "see" individual atoms and to manipulate individual atoms and molecules on various surfaces. One very interesting application of this technique is the construction of tiny "machines" made of atoms. A recent example of this activity was performed by a group of scientists from the University of Groningen in the Netherlands. They used carbon atoms to construct the tiny machine illustrated in the accompanying photo .... The scientists have been able to move the car forward as much as ten car lengths on the copper surface. (p. 63)

This textbook clearly points out that STM allows scientists to "see" and manipulate individual atoms and molecules to provide new materials that constitute nanotechnology (subject of Criterion #5). However, it is not made clear if the STM facilitates photographs or computer-generated images, and thus it was classified as Mention (M).

Following is an example of a textbook that was classified as Satisfactory (S), as it explicitly differentiates between computer-generated images and photographs: "The images in this box are *computer-generated representations, not true photographs*. However, they have opened our eyes to the appearance of surfaces in the most extraordinary ways" (Atkins & Jones, 2008, p. 189, italics added).

In a section entitled "Seeing Atoms," Oxtoby, Nachtrieb, and Freeman (1990) present the development of STM within a historical perspective:

> Microscopy began in the 15th century with the fabrication of magnifying glasses. By the late 17th century, the first optical microscopes were developed and used to observe single biological cells. *Optical microscopes are fundamentally limited*; the smallest things that can be distinguished with them have dimensions thousands of times the size of single atoms. In the 1930s, the invention of the electron microscope allowed scientists to bypass this limitation, and eventually this type of microscope was refined to the point that single atoms could be seen. The *disadvantage of the electron microscope* was that the high-energy electrons used in it very easily damaged samples ... scanning tunneling microscope ... uses an incredibly fine-pointed electrically conducting probe that is passed over the surface of the sample being examined .... When it comes nearly in contact with the atoms of the sample, a small electrical current (called the "tunneling current") can pass from the sample to the probe .... *The position of the probe is monitored and the information is stored in a computer* [and] a three-dimensional image of the surface can be constructed and displayed. (p. 29, italics added)

This textbook was classified as Satisfactory (S) for the following reasons: (a) Starting with magnifying glasses it provides an historical perspective for understanding STM; (b) Compares the disadvantages of optical and electron microscopes with respect to STM; (c) Explicitly states that in STM information is stored in a computer and images constructed and displayed.

One textbook after presenting details of STM, asked the following question: "How does the image obtained by a scanning tunneling microscope differ from that obtained by the usual optical microscope?" (McMurry & Fay, 2001, problem 2.24, p. 66). In a later section, the authors provided the following answer: "The image obtained with a scanning tunneling microscope is a three-dimensional, *computer-generated data plot* that uses tunneling current to mimic depth perception" (p. A-21, italics added). This presentation was classified as Satisfactory (S) as it clearly establishes a difference between the image of STM and an optical microscope.

According to Hill and Petrucci (1999):

> The wave-mechanical interpretation even includes the extremely small but nonzero possibility that an atom may transfer an electron to an adjacent atom without first ionizing. This can occur when an electron has a significant probability of being closer to the nucleus of another atom than to the nucleus of its "parent" atom. This transfer is called *tunneling* and requires much less energy than ionization .... A scanning tunneling microscope (STM) uses a tungsten probe with an extraordinary sharp tip that is carefully placed only about 0.5 nm from the surface being studied .... The flow of these electrons creates a small electric current .... The surface is scanned repeatedly, and the *results are processed by a computer to give a three-dimensional map of the surface.* These maps tell us how atoms are arranged on a surface. However, STM images tell us nothing about the internal structure of atoms. (p. 308)

This presentation was classified as Satisfactory (S) as it recognizes that STM maps are computer generated based on wave-mechanical properties of surface electrons and do not provide information about the internal structure of atoms. Following is another example of a presentation that was classified as Satisfactory (S):

> The probe tip [in STM] is moved systematically across the surface to form a complete *topographic map* of that part of the surface. The computer controlling the STM probe records the surface height at each location on the surface. These resulting topographic data are processed by software to form the final images that depict the surface contours. The STM image, which appears much like a photographic image, shows the locations of atoms on the surface being investigated. (The image is actually of the electrons on the atoms) (Moore, Stanitski, & Jurs, 2002, p. 49, italics added)

In a section entitled "Seeing Atoms" one of the textbooks stated: "The most modern way to see atoms is by use of special microscopes called scanning tunneling microscopes .... Such microscopes allow us to see computerized images of atoms by visualizing the electrical force fields around them. The currents are analyzed by computer and atomic images are displayed on computer monitors" (Dickson, 2000, p. 82). This presentation was classified as Satisfactory (S).

6.5.4 Criterion 4: Atomic Force Microscopy (AFM)

Ebbing and Gammon (2017) presented a fairly detailed account and following are some excerpts:

Both microscopes (STM and AFM) use a probe to scan a surface; but whereas the scanning tunneling microscope measures an electric current between the probe tip and the sample, the atomic force microscope measures the attractive van der Walls force between the probe tip and the sample. The advantage of the atomic force microscope is that it can be used with almost any surface, whereas the scanning tunneling microscope requires a conductive surface .... A computer coordinates the output from the photodiode with the sample position to create an image that appears on the computer screen (p. 862) .... [Next the authors provide an example of the type of surface that AFM can study] .... The image [Fig. 24.18] provided by an atomic force microscope is of rods of tobacco mosaic virus, a disease agent that infects tobacco and many other crops. Each virus rod is covered by molecules of bovine serum albumin (from the blood serum of cows). Chemists obtained this image as part of a study of the interaction of the albumin protein molecule with the virus (p. 862) (Classified as Satisfactory, S).

This is an interesting presentation (classified as Satisfactory) and following are some of its salient features: (a) Establishes a difference between STM and AFM; (b) Emphasizes that AFM can be used for most non-conductive surfaces; (c) The image produced is generated by a computer and not a photograph; and (d) An application of AFM in the resolution of an actual problem infecting crops.

Example of a textbook that was classified as Mention (M) is provided by the following presentation:

> The atomic force microscope (AFM), a modification of the scanning tunneling micro-
> scope, allows us to see groups of atoms. This is a single red blood cell [Figure 11–24].
> The AFM can also slice the cell to reveal individual protein molecules inside. (Cracolice
> & Peters, 2016, p. 308)

As compared to the previous presentation (Ebbing & Gammon, 2017), this was not classified as Satisfactory (S), as it does not clarify if the image is a photograph or generated by a computer.

These presentations are particularly helpful in understanding the difference between representation and presentation (intervention) that is manipulating to create new atom-sized products.

6.5.5   Criterion 5: From Representation to Presentation: Scientific Progress at a Crossroads

In a section entitled: are Atoms Real? McMurry and Fay (2001) stated that atomic theory lies at the heart of chemistry and then asked: how do we know that atoms are real? (p. 65). Authors then respond to this question in the following terms: "The best answer to that question is that we can now actually 'see' individual atoms with a remarkable device called a *scanning tunneling microscope*, or *STM* ... this special microscope has achieved magnifications of up to *10 million*, allowing chemists to look directly at individual atoms" (p. 65). It is plausible to suggest that the question of "are atoms real" approximates to what Daston and Galison (2007) have referred to as "representation" and the magnification of up to 10 million approximates "presentation." This presentation was classified as Mention (M).

According to Umland and Bellama (1999): "In an AFM, tip touches the sample but with a force low enough (about 10^{-9}N) that the surface is usually not damaged. The AFM maps the surface by 'feeling' interatomic forces similarly to the way blind people tap their canes to investigate the ground in front of them. Biological samples, such as chromosome clusters, can be observed under water, so that samples do not dry out, and even in vivo (in living systems). The motion associated with catalysis by an enzyme (a biological catalyst) was observed with an AFM in 1994" (p. 469). Classified as Mention (M). Another textbook referred to individual atoms be seen as bumps on the surface of a solid by the technique called scanning tunneling microscope, STM (Atkins & Jones, 2008, p. F16). Following is another example of a textbook that was classified as Mention (M): "In this image made by scanning tunneling microscope (STM) individual silicon atoms on the surface of a silicon crystal are seen at a magnification of 10 million" (Hill & Petrucci, 1999, p. 308). Once again, it is important to note that magnification can be understood as "presentation." One of the textbooks did not refer to STM or AFM and still provided the following statement with respect to "nanoworld":

> Consumer products containing materials produced by nanotechnology began showing up in the mid-1990s. Two common applications at that time were the inclusion of nanometer-sized particles (called nanoparticles) in cosmetics and sunscreen products …. The sport of tennis has benefitted significantly from nanotechnology. One company injects nanoparticles of silicon dioxide into voids in the graphite frame of their tennis rackets. The result is a stronger frame that allows more power to be delivered to the ball with each stroke …. Nanoparticle-based textile treatments have revolutionized the textile industry by making possible products such as quick-drying, waterproof, wrinkle-free, and stain-resistant clothing (Seager & Slabaugh, 2011, p. 79). Classified as Mention (M).

Another textbook also did not refer to STM or AFM and provided the following statement with respect to "nanotubes":

> New variations on the fullerenes are nanotubes …. Nanotubes come in a variety of forms. Single-walled carbon nanotubes can vary in diameter from 1 to 3 nm and are about 20 nm long. These compounds have generated great industrial interest because of their optical and electrical properties. They may play a role in miniaturization of instruments, giving rise to a *new generation of nanoscale devices*. (Bettelheim, Brown, Campbell, & Farrell, 2010, p. 164, italics added)

The reference to a new generation of nanoscale devices clearly shows the role played by nanotechnology in the transition from "representation" to "presentation" and hence scientific progress at a crossroads (Daston & Galison, 2007; Hoffmann, 2012).

Following are examples of eight textbooks that were classified as Satisfactory (S). Some of these presentations are discussed later to highlight their salient features and provide an overview of how we are at a crossroads:

> Nanotechnology, the field of trying to build ultrasmall structures one at a time, has progressed in recent years. One potential application of nanotechnology is the construction of artificial cells. The simplest cells would probably mimic red blood cells, the body's oxygen transporters. For example, nanocontainers, perhaps constructed of carbon, could be pumped full of oxygen and injected into a person's bloodstream. If the person needed additional oxygen—due to heart attack perhaps, or for the purpose of space travel—these containers could slowly release oxygen into the blood, allowing tissues that would otherwise die to remain alive. (Tro, 2008, p. 42, as part of a section entitled "Challenge Problems")

A new area of research with the potential, among other things, for revolutionizing medical diagnosis and treatment and improving our quality of life is **nanoscience**. This field includes the study of materials that are larger than single atoms, but too small to exhibit most bulk properties .... **Nanomaterials** are materials composed of nanoparticles or regular arrays of molecules or atoms such as nanotubes (Box 14.1). These materials have been made possible by advances in nanotechnology, such as new imaging technologies, including the scanning tunneling microscope (see Box 5.1), and the discovery of how some non-metals and metalloids can be manipulated into assembling themselves into regular, extended, structures. (Atkins & Jones, 2008, p. 648, original emphasis)

Metals also have unusual properties on the 1–100 nm-length scale. Fundamentally this is because the mean free path of an electron in a metal at room temperature is typically about 1–100 nm. So when the particle size of a metal is 100 nm or less, one might expect unusual effects .... Other physical and chemical properties of metallic nanoparticles are also different from the properties of the bulk materials. Gold particles less than 20 nm in diameter melt at a far lower temperature than bulk gold, for instance, and when the particles are between 2 and 3 nm in diameter, gold is no longer a "noble," unreactive metal; in this size range it becomes chemically reactive. At nanoscale dimensions, silver has properties analogous to that of gold in its beautiful colors, although it is more reactive than gold. Currently, there is great interest in research laboratories around the world in taking advantage of the unusual optical properties of metal nanoparticles for application in biomedical imaging and chemical detection. (Brown, Le May, Bursten, Murphy, & Woodward, 2014, p. 554, in a section entitled "Metals on the nanoscale")

Through the use of scanning tunneling microscopy ... pioneers of nanotechnology believe that, by arranging structures one atom at a time, they can create simple computers the size of bacteria or computers a million times more powerful than today's desktop models the size of a sugar cube! Medical devices could be made so precise that individual cells, even individual genes, could be targeted surgically or pharmacologically (p. 469) .... [In another section this textbook refers to nanotubes] These younger cousins of fullerenes [C_{60} structure represents a third form of crystalline carbon, graphite and diamond being the other two] consist of extremely long, thin, graphite-like cylinders with fullerene ends. They are often nested within one another .... Despite thickness of a few nanometers, *these structures are highly conductive along their length and about 40 times stronger than steel! Dreams abound of nanoscale electronic components made with nanotubes* or atom-thick wires formed by inserting metal atoms within the interior. (Silberberg, 2000, p. 383, italics added)

Carbon nanotubes conduct electricity because, like conducting polymers, they have an extended network of delocalized π-bonds. Electrons are delocalized from one end of the tube to the other. Along the long axis of the tube, the conductivity of a carbon nanotube can be high enough to be considered metallic .... Hydrogen absorbed into nanotubes can be stored in a volume much smaller than that required to store the gas. (Jones & Atkins, 2000, p. 871)

Can you imagine a thermometer that has a diameter equal to one one-hundredth of a human hair? Such a device has actually been produced by scientists Yihua Gao and Yoshio Bando of the National Institute for Materials Science in Tsukuba, Japan. The thermometer they constructed is so tiny that it must be read using a powerful electron microscope. It turns out that the tiny thermometers were produced by accident. The Japanese scientists were actually trying to make tiny (nanoscale) gallium nitride wires. However, when they examined the results of their experiments, they discovered tiny tubes of carbon atoms that were filled with elementary gallium. Because gallium is a liquid over an unusually large temperature range, it makes a perfect working liquid for a thermometer ... gallium moves up the tube as the temperature increases. These minuscule thermometers are not useful in the normal macroscopic world—they can't be seen with the naked eye. However, they should be valuable for monitoring temperatures from 50°C to 500°C in materials in the nanoscale world. (Zumdahl & DeCoste, 2015, p. 35, in a section entitled, "Tiny Thermometers")

Robert A. Wolkow and co-workers at the Canadian National Institute for Nanotechnology at the University of Alberta have developed a technique for a silicon surface with single-atom-thick layer of hydrogen atoms. They can then selectively remove one or more hydrogen atoms, leaving negatively charged silicon atoms at the surface. When one hydrogen atom is removed, the single, negatively charged silicon atom at the surface behaves as a quantum dot .... Two electrons in this four-atom group can be manipulated by removing other hydrogen atoms .... One arrangement of the two electrons can be considered "switch off" and the other (diagonal to the first) can be considered "switch on." Techniques exist by which such switches could be made to behave as a circuit for a computer, thereby allowing much smaller computers and thinner cell phones than ever before. (Moore, Stanitski, & Jurs, 2011, p. 21, in a section entitled "Atomic scale electric switches")

The interior of a buckyball is large enough to hold an atom of any element in the periodic table. Researchers wasted no time in putting different metals' atoms in the center of buckyballs. Thus, the result was a new family of superconductors. Other teams are working on using buckyballs as the source of tiny ball bearings, lightweight batteries, and even super-conducting wires that are just one-cluster thick. (Cracolice & Peters, 2016, p. 432)

The presentation by Brown et al. (2014) can be of particular interest to students as general chemistry textbooks present silver and gold as the unreactive "noble" metals. However, at the nanoscale the manipulation of the size of the particles facilitates reactivity with useful biomedical and chemical applications. Furthermore, these authors provide a historical backdrop to this application by pointing out that even in the fifteenth century the artisans who prepared stained-glass knew that gold when dispersed in molten glass acquired a beautiful profound red color. Of course these artisans were not aware of the underlying complexity of the processes involved. As an example they reproduce a stained-glass window from Milan's cathedral in beautiful colors. Again, the authors provide another example of colloidal solutions of small particles of gold (with intense red color) prepared by Michael Faraday around 1857, which are still conserved in the Faraday Museum in London. Similarly, Brown, LeMay, Bursten, and Murphy (2009) also provide a very similar presentation along with the historical examples (Milan's cathedral and the Faraday Museum). Interestingly, an earlier edition of this textbook (Brown, LeMay, & Bursten, 1997) does not provide similar details. This shows how textbooks can incorporate new material over time, in new editions.

Silberberg (2000) refers to dreams of nanoscale materials that are highly conductive and 40 times stronger than steel. Indeed, such dream-like materials can provide students an opportunity to go and look beyond our present state of scientific progress to be at a crossroads. Jones and Atkins (2000) suggest that hydrogen absorbed into nanotubes would solve a major obstacle to the use of hydrogen fuel cells, by providing a compact storage medium.

The presentation by Zumdahl and DeCoste (2015) not only introduce students to new products, "Tiny Thermometers" (tubes of carbon atoms filled with gallium) at the nanoscale but also suggest how they can in the future replace what actually exists (mercury thermometers). Furthermore, it also shows the role played by chance discovery as the Japanese scientists who built these thermometers were actually working on gallium nitride wires as part of nanotechnology.

Moore, Stanitski, and Jurs (2011) not only provide an example of how a device made by nanotechnology (electric switches) can be made and useful in the construction of new computers and cell phones, but also provide a reference to the laboratory where the research is being conducted (Haidar, Pitters, Di Labio, Livadark, Mutus, & Wolkow, 2009). Furthermore, another example of nanotechnology is provided by showing how when water freezes on a copper surface the water molecules are arranged in a pentagonal pattern, instead of the usual hexagonal (p. 407). Such information can provide students a real life experience of scientists who are actually working in cutting-edge research and thus provide an incentive for doing further research.

Brady and Senese (2009) not only provide a fairly good description of nanotechnology, but have also included on the cover of the textbook a carbon nanotube emerging from glowing plasma of hydrogen and carbon. Furthermore, they have included an exercise of critical thinking: "Graphite is a reasonably good conductor in directions parallel to the planes of the carbon atoms, but is a poor conductor in a direction perpendicular to the planes. Why is this so? Would you expect carbon nanotubes to be good conductors of electricity along their length?" (Brady & Senese, 2009, p. 899).

The presentation by Cracolice and Peters (2016) is particularly helpful in understanding the versatility of the materials that can be produced and their properties manipulated by introducing different atoms into the buckyballs.

Brown and Holme (2011) followed a novel approach by pointing out how science keeps progressing even if you think that everything is already known about a particular topic: "When you think about the elements in the periodic table, you probably assume that most things are known about them. For decades, chemistry textbooks said that there were two forms of the element carbon: graphite and diamond. In 1985, that picture changed overnight. A team of chemists at Rice University discovered a new form of carbon, whose 60 atoms formed a framework that looks like a tiny soccer ball. Because the structure resembled the geodesic domes popularized by the architect Buck minster Fuller, it was given the whimsical name of buckminsterfullerene" (p. 241). Actually, not only students but even many chemists had perhaps not envisioned a third form of carbon, and this brief introduction to the origin of nanotechnology can open a whole new perspective for students who are preparing for various types of careers. With this background authors provide the following example of a nanomaterial and its application in the drug industry:

> A promising method for this type of drug delivery system uses a material called mesoporous silica nanoparticles (MSN). As we saw earlier, silica is composed of net-works of SiO_4 units, and those SiO_4 units can form a honeycomb structure as shown in Figure 7.18. As MSN is simply a very small particle with this honeycomb arrangement, because of this structure, these particles have enormous surface to volume ratios: one gram of the material has roughly the same surface as a football field. Once loaded with

the desired therapeutic agents, the pore can be capped with another molecule, and the whole nanoparticle is delivered to the target (Brown & Holme, 2011, p. 234). This presentation was classified as Satisfactory (S).

The idea of "surface to volume ratio" can be particularly helpful in understanding the degree to which nanomaterials can be manipulated. Furthermore, authors point out that amorphous silica particles can destroy red blood cells and so are not biocompatible. However, surprisingly MSN are biocompatible.

Spencer, Bodner, and Rickard (2012) highlighted the following aspects of nanomaterials:

> C_{60} is now known to be a member of a family of compounds known as fullerenes. C_{60} may be the most important of the fullerenes because it is the most perfectly symmetrical molecule possible, spinning in the solid state at the rate of more than 100 million times per second. Because of their symmetry, C_{60} molecules pack as regularly as ping-pong balls. The resulting solid has unusual properties. Initially, it is soft as graphite, but when compressed by 30%, it becomes harder than diamond. When this pressure is released, the solid springs back to its original volume. C_{60} therefore has the remarkable property that it bounces back when shot at a metal at high speeds. (p. 373)

These remarkable properties of the fullerenes clearly show the importance of manipulating nanomaterials to provide entirely new types of materials that can be suitable for different purposes in the industry. This presentation was classified as Satisfactory (S).

Olmsted and Williams (2006) not only describe the applications of nanotechnology but also discuss the difficulties involved in constructing nanomaterials and how these are being resolved:

> Supposing that scientists succeed in constructing molecular tools, they must overcome another obstacle for nanotechnology to be effective. A medical nanosubmarine is likely to contain about a billion (10^9) atoms. At an assembly speed of one atom per second, it would take 10^9 seconds to construct one such device. That's almost 32 years! If the assembly rate can increased to one atom per microsecond, the construction time for a 1-billion-atom machine drops to 1000 seconds, or just under 17 minutes .... To be practiced, then, nanotechnology must be precise, extremely fast, and amenable to mass production. Perhaps this strikes you as definitely in the realm of science fiction rather than science fact, and perhaps it is. Nevertheless, scientists at many universities are vigorously tackling the challenges of this field, and major technology companies have active research groups as well (p. 37). Classified as Satisfactory (S).

Indeed, not only science students, but even teachers and scientists themselves are amazed at the possibilities being offered by nanotechnology.

Finally, it is important to note that as suggested by Hoffmann (2016b), although the tension between representation and presentation has always been recognized in chemistry as a science, perhaps the same does not hold for science (chemistry) education. This study shows that only 25% of the general chemistry textbooks (published in USA) evaluated had a satisfactory presentation with respect to the importance of nanotechnology (see Table 6.1, Criterion 5). However, some caution is necessary in interpreting recent developments in nanotechnology. In order to receive further feedback I sent this version of Chap. 6 to Roald Hoffman, who responded in the following terms: "I have now had time to read your Chap. [6]. It has a great summary of my views and those of Galison in the beginning, focusing on the work of Hacking both of us refer to. And a very good analysis of the way the scientific method and STM/AFM are treated in textbooks .... The reference to the Millikan experiments are excellent" (Hoffmann, R., Email to author, July 23, 2016c).

Research reported in this chapter has shown the importance of transgression versus objectivity and nanotechnology and how these subjects are dealt with in general chemistry textbooks. Conclusions based on these findings will be integrated with those from other chapters and presented in Chap. 7.

References

Binns, I. C., & Bell, R. L. (2015). Representation of scientific methodology in secondary science textbooks. *Science & Education, 24*, 913—936.

Brito, A., Rodríguez, M.A., & Niaz, M. (2005). A reconstruction of development of the periodic table based on history and philosophy of science and its implications for general chemistry textbooks. *Journal of Research in Science Teaching, 42*, 84–111.

Chalmers, A. (2009). *The scientist's atom and the philosopher's stone: how science succeeded and philosophy failed to gain knowledge of atoms.* Dordrecht: Springer.

Cortéz, R., & Niaz, M. (1999). Adolescents' understanding of observation, prediction, hypothesis in everyday and educational contexts. *Journal of Genetic Psychology, 160*(2), 125–141.

Daston, L., & Galison, P. L. (1992). The image of objectivity. *Representations, 40*(special issue: Seeing Science), 81–128.

Daston, L., & Galison, P. (2007). *Objectivity.* New York: Zone Books.

De Berg, K. C. (2011). Joseph Priestley across theology, education, and chemistry: an interdisciplinary case study in epistemology with a focus on the science education context. *Science & Education, 20*(7–8), 805–830.

Denzin, N. K., & Lincoln, Y. S. (2005). Introduction: the discipline and practice of qualitative research. In N. K. Denzin & Y. S. Lincoln (Eds.), *The Sage handbook of qualitative research.* 3rd ed. (pp. 1–32). Thousand Oaks: Sage.

Duhem, P. (1914). The aim and structure of physical theory (2nd ed., trans: Wiener, P. P.). New York: Atheneum.

Feyerabend, P. (1975). *Against method: outline of an anarchist theory of knowledge.* London: New Left Books.

Grandy, R., & Duschl, R. A. (2007). Reconsidering the character and role of inquiry in school science: analysis of a conference. *Science & Education, 16*, 141–166.

Hacking, I. (1983). *Representing and intervening.* Cambridge: Cambridge University Press.

Hacking, I. (1984). Experimentation and scientific realism. In J. Leplin (Ed.), *Scientific realism.* Berkeley: University of California Press.

Haidar, M. B., Pitters, J. L., Di Labio, J. L., Livadark, L., Mutus, J. Y., & Wolkow, R. A. (2009). Controlled coupling and occupation of silicon atomic dots at room temperature. *Physical Review Letters, 102*, 046805.

Hodson, D. (2009). *). Teaching and learning about science: language, theories, methods, history, traditions and values.* Rotterdam: Sense Publishers.

Hoffmann, R. (2006). Images from the nanoworld challenge viewers' thinking. *American Scientist, 94*, 1–5.

Hoffmann, R. (2012). J. Kovac & M. Weisberg (Eds.), What might philosophy of science look like if chemists built it? *Roald Hoffmann on the philosophy, art, and science of chemistry.* (pp. 21–38). New York: Oxford University Press.

Hoffmann, R. (2014). The tensions of scientific storytelling: science depends on compelling narratives. *American Scientist, 102*, 250–253.

Holton, G. (1978a). Subelectrons, presuppositions, and the Millikan-Ehrenhaft dispute. *Historical Studies in the Physical Sciences, 9*, 161–224.

Holton, G. (1978b). *The scientific imagination: case studies.* Cambridge: Cambridge University Press.

Holton, G. (2014b). Email to author, dated August 3.

Lakatos, I. (1970). Falsification and the methodology of scientific research programs. In I. Lakatos & A. Musgrave (eds.), *Criticism and the growth of knowledge* (pp. 91–195). Cambridge: Cambridge University Press.

Levere, T. H. (2006). What history can teach us about science: theory and experiment, data and evidence. *Interchange, 37*, 115–128.

Losee, J. (2001). *A historical introduction to the philosophy of science* 4th ed. Oxford, UK: Oxford University Press.

Matthews, M. R. (1992). History, philosophy and science teaching: The present reapproachment. *Science & Education, 1*(1), 11–47.

Millikan, R. A. (1913). On the elementary electrical charge and the Avogadro constant. *Physical Review, 2*, 109–143.

Motterlini, M. (1999). *For and against method: including Lakatos's lectures on scientific method and the Lakatos-Feyerabend correspondence.* London: University of Chicago Press.

Needham, P. (2004). Has Daltonian atomism provided chemistry with any explanations? *Philosophy of Science, 71*, 1038–1047.

Niaz, M. (1998). From cathode rays to alpha particles to quantum of action: a rational reconstruction of structure of the atom and its implications for chemistry textbooks. *Science Education, 82*, 527–552.

Niaz, M. (2000). The oil drop experiment: a rational reconstruction of the Millikan-Ehrenhaft controversy and its implications for chemistry textbooks. *Journal of Research in Science Teaching, 37*, 480–508.

Niaz, M. (2009). *Critical appraisal of physical science as a human enterprise: dynamics of scientific progress.* Dordrecht: Springer.

Niaz, M. (2011). *Innovating science teacher education: a history and philosophy of science perspective.* New York: Routledge.

Niaz, M. (2012). *From 'Science in the Making' to understanding the nature of science: an overview for science educators.* New York: Routledge.

Niaz, M. (2016). *Chemistry education and contributions from history and philosophy of science.* Dordrecht: Springer.

Niaz, M., & Maza, A. (2011). *Nature of science in general chemistry textbooks.* Dordrecht: SpringerBriefs in Education.

Niaz, M., & Rivas, M. (2016). *Students' understanding of research methodology in the context of dynamics of scientific progress.* Dordrecht: SpringerBriefs in Education.

Olenick, R. P., Apostol, T. M., & Goodstein, D. L. (1985). *Beyond the mechanical universe: from electricity to modern physics.* New York: Cambridge University Press.

Perl, M. L., & Lee, E. R. (1997). The search for elementary particles with fractional electric charge and the philosophy of speculative experiments. *American Journal of Physics, 65*, 698–706.

Preston, J. (1997). *Feyerabend: Philosophy, science and society.* Cambridge, UK: Polity Press.

Rocke, A. (2013). What did 'theory' mean to nineteenth-century chemists? *Foundations of Chemistry, 15*, 145–156.

Tiles, J. E. (1994). Experiment as intervention. *British Journal for the Philosophy of Science, 44* (3), 463–475.

Windschitl, M. (2004). Folk theories of "inquiry": how preservice teachers reproduce the discourse and practices of an atheoretical scientific method. *Journal of Research in Science Teaching, 41*, 481–512.

Ziman, J. (1978). *Reliable knowledge: an exploration of the grounds for belief in science.* Cambridge: Cambridge University Press.

Chapter 7
Conclusion: Understanding the Elusive Nature of Objectivity

An evaluation of research in science education reported in this book shows the problematic nature of understanding some of the universal values associated with objectivity such as certainty, value neutral observations, facts, infallibility, scientific method, and truth of scientific theories and laws. Similarly, aspects of Merton's "ethos of science" such as open-mindedness, universalist, disinterested, and communal have also been invoked to understand progress in science. Studies evaluated, however, have pointed out that some of these values are not necessarily essential for understanding objectivity. Philosophy of science itself has explored new territory in this context and Giere (2006a, p. 95) considers that it is *presentist hubris* to think that we can have an objectively correct or true theories. Daston and Galison (2007) have constructed the evolving nature of scientific judgment (objectivity) through the following phases: truth-to-nature, mechanical objectivity, structural objectivity, and finally trained judgment. Each of these regimes did not supplant the other but they can coexist and supplement each other at the same time. Although objectivity is not synonymous with truth or certainty, it has eclipsed other epistemic virtues and to be objective is often used as a synonym for scientific in both science and science education.

Table 7.1 provides an overview of the classification of all the articles evaluated in this book. Following are some of the salient features of the results obtained: (a) S&E was the only journal in which two articles were classified in Level V, which approximates to the evolving nature of objectivity; (b) In all the chapters most of the articles were classified in Levels II and III; (c) Classification of the articles in Level III (62% for JRST, and 44% for S&E) means that the authors recognized the problematic nature of objectivity and hence the need for alternatives; and (d) Very few articles were classified in Level I (none for HPST and ESE), which approximates to the traditional concept of objectivity as found in most science textbooks. These results provide a detailed account (over a period of almost 25 years) of how the science education research community conceptualizes the difficulties involved in accepting objectivity as an unquestioned epistemic virtue of the

© Springer International Publishing AG 2018 179
M. Niaz, *Evolving Nature of Objectivity in the History of Science and its Implications for Science Education*, Contemporary Trends and Issues in Science Education,
DOI 10.1007/978-3-319-67726-2_7

Table 7.1 Comparison of the levels of classification of articles in different chapters of this book

Chapter (Journal)	n	No. of articles classified in level				
		I	II	III	IV	V
3 (S&E)	131	5	56	58	10	2
4 (JRST)	110	4	33	68	5	–
5 (HPST)	8	–	4	3	1	–
5 (ESE)	12	–	6	4	2	–

Notes:
1. For a description of levels I–V see Chap. 3
2. n: Total number of articles evaluated
3. S&E: *Science & Education*
4. JRST: *Journal of Research in Science Teaching*
5. HPST: *International Handbook of Research in History, Philosophy & Science Teaching*
6. ESE: *Encyclopedia of Science Education*

scientific enterprise. Nevertheless, it seems that more work needs to be done in order to facilitate a transition (Levels IV and V) toward a more nuanced understanding of objectivity and eventually the dynamics of scientific progress.

Following are some aspects for facilitating an understanding of the elusive nature of objectivity based on articles evaluated in S&E, JRST, HPST, and ESE. Furthermore, based on the evaluation of general chemistry textbooks the idea of "transgression of objectivity" is introduced. It is plausible to suggest that these findings have implications for science education which are synthesized and discussed in the following sections (Based on Chaps. 3, 4, 5, and 6 and presented in alphabetical order):

7.1 Alternative Interpretations of Data in Science and Objectivity

Science textbooks generally expound on a series of theories that deal with a topic and it is tacitly understood that theoretical and methodological standards for selecting a theory are neutral and objective. However, what is missing is an essential aspect of scientific progress namely based on alternative theoretical frameworks the same data can be interpreted differently by scientists before reaching consensus with respect to the canonical nature of science. Reproducibility of scientific experiments is generally considered to contribute to the objectivity of scientific knowledge. However, this ignores the difficulties faced by students and even scientists to reproduce and interpret experimental results. Thus, the subjectivity involved in different interpretations is necessary for understanding the dynamics of "science in the making." Alternative interpretations of data provide students the opportunity to understand that progress in science involves rivalries and uncertainty, which precisely leads to controversies among scientists.

7.2 Alternative Research Methodologies and Objectivity

Scientists in different disciplines have distinct epistemic goals and practices and consequently their conceptions of rationality and objectivity can also vary. However, science is generally portrayed as a source of objective knowledge and a possible contribution of scientific narratives based on students' thinking is considered to be subjective and thus ignored. An example of alternative research methodologies is the mixed methods research based on interviews (among other methodologies) with the students that can facilitate more positive attitudes toward science. In a sense this corroborates what scientists themselves do by interacting within the scientific community (i.e., trained judgment as suggested by Daston and Galison, 2007). It is plausible to suggest that the objectivity of science (and also mathematics) rests on the criticizability of the different arguments put forward by the scientists.

7.3 Canonizing Objectivity to Reinforce Privileges

Academic achievement gap between different sectors of a society is a cause for concern (e.g., African American and white students in the USA). Dominant groups in a culture generally support the existing structure of objective knowledge and any attempt to question it is considered as opposition and insubordination. This leads the dominant group to consider its understanding as the canonized version of objectivity. Similarly, science also provided the objective evidence of the natural inferiority of women, the homosexuals, the colonized, and the enslaved. Furthermore, science is often envisioned as directly reflecting the truths in science and therefore unquestionable. Diversity of views helps to understand objectivity and it is undermined if the objective correctness of a claim is taken to be what is endorsed by a privileged point of view.

7.4 Empiricist Epistemology and Objectivity

School science generally emphasizes an empiricist epistemology in which a "purified" version of science is considered essential for achieving the "unobtainable" ideals of truth and objectivity. In the late nineteenth century the manipulation of physical objects and instruments helped to reframe mathematics and astronomy as a physical science, which facilitated a culture of objectivity. Such practice leads to a "myth of experimenticism" namely following the path from experiment to theory (emphasis on empirical methods) not only does not provide greater objectivity but also deprives students of an environment that facilitates thinking and understanding arguments. Contrary to popular belief in science education, simple Baconian stockpiling and ordering of observations does not facilitate the formation of better scientists.

7.5   Femininity-Masculinity, Science and Objectivity

According to feminist critics (Harding, Keller, and Longino) science has grown out of a Western male tradition that celebrates objectivity and power relations based on masculinity that leads to dualisms such as: rational-emotional, logical-intuitive, objective-subjective, and abstracted-holistic. Both science and science education assert the relationship between masculinity and traits such as objectivity, rationality, and lack of emotion. Furthermore, if femininity is viewed as mutually exclusive with masculinity, this also leads to femininity being considered as lacking the scientific traits. The politically engaged standpoint of feminism is less partial and distorted than the standpoint of conventional scientific inquiry. Overcoming such simplistic relationships and dualisms can facilitate a critical examination of science and a better understanding of gender in science education. This conflict between masculine and feminist traits can at times lead women to abandon careers in science. Similarly, the abstraction and objectivity of pure science have masculine connotations, whereas the human and social sciences are considered to be feminine.

7.6   Interaction Between Evidence and Belief (Faith) and the Quest for Objectivity

Interactions between evidence and faith become important in controversial issues such as teaching evolution in a biology course. The dilemma faced by the teacher is based on the fact that although students may seem to understand evolution (based on evidence) they generally do not believe in it. In such cases, the cultural milieu of the students in which the subject is taught is important. It is even suggested that today's teacher of evolution faces a situation very similar to Darwin when he presented the *Origin of Species*. Consequently, the interaction between the understanding based on belief in the absence of objective evidence and acceptance based on evidence can provide a better understanding of the nature of science. In general, teachers' perceptions and beliefs about learning also affect how they approach the material and what they teach. At this stage a word of caution is necessary: if our goal as teachers is to get students to believe the content we teach—then that may be considered as indoctrination and not education (I owe this observation to Aikenhead, 2016). More recently, based on contemporary epistemology/philosophy of mind scholarship, Smith and Siegel (2016) have clarified that belief is involuntary and need not be used as a basis for inference or action, whereas acceptance is voluntary and involves a commitment to use what is accepted in one's practical reasoning.

However, it is important to note a caveat based on history of science, which shows that in some cases even scientists do not agree with respect to the interpretation of evidence as they have different prior epistemological beliefs. In other words, changes in science can occur by means other than rational consideration of empirical evidence. Furthermore, although objectivity is a value that all scientists strive for in their work, what is a fact in science is continually reevaluated in the light of ongoing research.

7.7   Mertonian "Ethos of Science" and Objectivity

Merton's (1942) "ethos of science" is based on norms of universalism, commun-ism, disinterestedness, and organized skepticism. Merton believed that these insti-tutional values are transmitted by precept and example, perhaps during the course of a scientist's educational career and can even be considered as the idealized "view from nowhere" (Reiss, 2014). Is there a contradiction between Merton's "ethos of science" and Daston and Galison's understanding of objectivity in science? It seems that increasing commodification may jeopardize Mertonian norms of openness in scientific practice, truthfulness, objectivity, trust, accuracy, and respect for expertise (Vermeir, 2013). Similarly, social constructivism may jeopardize Merton's "ethos of science" (Slezak, 1994). Longino (1990) under-scores the need for criticism from alternative perspectives and thus postulates a social structure for achieving Merton's "organized criticism." Merton's universal-ism does seem to imply the objectivity of scientific knowledge (McCarthy, 2014). However, if we do not conflate objectivity with universal and unconditional cor-rectness of scientific knowledge, but rather consider scientific inquiry to provide a greater degree of objectivity (Daston & Galison, 2007), then Merton's ethos of science can still provide guidance. Despite these difficulties, it seems that Mertonian norms of the scientific enterprise can be reinforced by following the process of trained judgment rather than mechanical objectivity.

7.8   Objectivity as a Process and not a State

Understanding of constructivism in Piaget's theory of cognitive development and genetic epistemology has been the subject of considerable controversy in science education research. In this theoretical framework, construction of knowledge by the child is the result of a subjective knower within a social context that facilitates transformation, organization, and interpretation of structures leading to a dialecti-cal interaction. An important implication for science education is that "objectivity is a process and not a state" (Piaget, 1971) that means a continuous series of suc-cessive approximations toward objectivity that may never be achieved. In other words, objectivity is not an "all or nothing thing," but rather it comes in degrees (Machamer & Wolters, 2004). Similarly, in Piaget's genetic epistemology, study-ing the psychological subject can lead to an approximation toward the epistemic subject. This means that a classroom teacher needs to be more concerned about the process (and not the product) that can facilitate an approximation toward what may be considered as objective or even perhaps iconic knowledge in a particular domain of science content. Research reported in science education provides evidence for constructivist teaching strategies that facilitate change by taking into consideration students' alternative conceptions as part of the process of conceptual understanding. Such experiences lead to innovative teaching strategies based on the following: audit the process rather than the product. Similarly, Gergen's (1994) understanding of objectivity as involving the dynamics of process-product complements Daston and Galison's (1992) truth-to-nature. It is plausible to suggest that the underlying ideas of Daston and Galison, Gergen and Piaget (formulated in different domains of knowledge) go beyond the positivist understanding of objectivity and even complement each other. Similarly, it seems that there is a possible relationship between Cushing's (1995) idea of *contingency* and the historical evolution of the regime of objectivity as presented by Daston and Galison (2007). Cushing refers to the hegemony of the Copenhagen interpretation of quantum mechanics over its rivals on non-epistemic reasons, that is on grounds that were not necessarily objective or rational. In Daston and Galison's (2007) framework this could be understood as an episode in which trained judgment of the community prevailed. It is plausible to suggest that it is perhaps the contingent nature of science (Cushing, 1995) that manifests itself in the evolving nature of objectivity. Quantum mechanics and valence theory provide good examples of such changing or competing theories (cf. Niaz, 2016).

7.9   Objectivity and Value Neutrality in Science

As scientists are part of a society, the notion that a scientific expert can be entirely neutral, value-free, and objective is difficult to understand. Despite efforts to present science as objective and autonomous, its relationship with capital and market forces is well known, and at times a picture of value-free science is presented as more of a romantic principle. The argument for a value-free science is difficult to sustain as most human activities are value-laden and historians of science have recognized this facet of the progress in science. Consequently, although historians, philosophers of science, and science educators may aspire for a science that is value-free, neutral and objective, the real picture of the scientific enterprise is much more complex. Furthermore, insisting on the objectivity and neutrality of science and ignoring the social forces that determine its progress does not facilitate a critical appreciation by students. Following are some examples of topics that involve ethics and values: depletion of ozone layer, genetics, gene therapy, stem cell research, xenotransplantation, napalm, agent orange, pollution, nuclear weapons, garbage collection, among others. Furthermore, history of science shows that facts have rarely been loyal to values which initially led to their identification. A good example is Darwin's use of facts that had been gathered by his teleologically oriented predecessors associated with a different set of values. In the case of gender and phrenology, objectivity and neutrality of the scientific enterprise was compromised. Consideration of the problematic nature of value neutrality leads to a thought-provoking question: if the ideal of value-free inquiry is flawed, what is to replace it? Based on Longino (2002), a possible alternative is "social value management" which involves non-epistemic values (social, economic, and other) (Irzik, 2015). As science is a human construction, scientists first prefer their own interpretation of the data, which may or may not change (or change partially) under the scrutinizing lenses of the scientific community. In this context, it would be interesting if courage, humility, and willingness to suspend judgment can also be considered as necessary values in the scientific enterprise. Holton (1978a, b) has, for example, recognized the role of "willingness to suspend judgement" in the historical reconstruction of the oil drop experiment. It is plausible to suggest that there is an underlying tension between scientific progress and the assumptions with respect to its neutrality and objectivity.

7.10   Objectivity-Subjectivity as the Two Poles of a Continuum

In most educational systems, the virtues of the traditional scientific tradition (rationality, objectivity, and skepticism) are challenged by strands of irrationality, subjectivity, and credulity and this can pose considerable problems for the science teacher. However, such dual ways of thinking also formed part of the progress of science itself. Precisely, the evolving nature of objectivity based on the history of science can be a source of guidance for the educational community.

Quantitative research methodology in education can be associated with positivistic styles of thinking. On the other hand, integration of qualitative and quantitative research methodologies can provide a better understanding of objectivity by facilitating competition between divergent approaches to research. In the interpretive research paradigm (social constructivist), traditional standards of internal and external validity, reliability and objectivity are replaced or complemented with notions of credibility, transferability, dependability, and confirmability. Triangulation based on different data sources is particularly helpful in enhancing credibility of the research. Such research experiences inevitably recognize the relationship between objectivity and the underlying subjectivity that leads to the creation of multiple realities. The dualism between objectivity and subjectivity leads to a conflict in the evolving nature of progress in science. Ignoring this duality may lead to the hegemony of objective knowledge and the consequent emphasis on rote learning. During scientific research, subjective and objective aspects interact by means of communication and peer reviews within the scientific community. In the case of students' thinking of nature of science, it is plausible to suggest that it progresses from one pole of empiricist epistemology to another, which considers subjective limitations in both components of scientific knowledge, namely empirical evidence and coordinating theory. It has also been argued that in qualitative research a detached observer claiming objectivity would not be able to access suitable data. Daston and Galison (1992, p. 82) have referred to a similar tension between subjectivity and objectivity, in the history of science itself. In other words, the personal construction of the students (subjective) can always be contrasted with the objective canonical knowledge, leading to integration. Those who work in the lab (both students and scientists) can face a dilemma when they have to deal with the subjectivist doubts with respect to observations. It is plausible to suggest that "trained judgment" could be one alternative to reach consensus with respect to the interpretation of observed data.

At present there is considerable debate in science education with respect to assessment of students' performance based on multiple-choice questions (considered objective) and conceptual problems (considered subjective). Despite this debate, the research community also recognizes that multiple-choice questions are generally based on memorized algorithms and do not facilitate meaningful learning of science content. The tension between subjectivity and objectivity in assessment provides an opportunity to reflect upon the very essence of the scientific enterprise, namely doing and understanding science involves interpretation and not just memorizing algorithms, hence the importance of conceptual problems. Furthermore, school science is generally considered to be *scientific* that is characterized by rationality, precision, formality, detachment, and objectivity. In contrast, *everyday* science is considered to denote an opposing set of characteristics such as improvisation, ambiguity, informality, engagement, and subjectivity. It can be argued that the two sets of characteristics are not dichotomous but change continuously depending on the needs of the school environment.

In controversial topics of the science curriculum such as evolution, the instructor with a professional training in evolutionary biology thinks that he is being objective, and still at the same time in his interactions with the students he/she is forced to grapple with issues that require subjective understanding. This once again illustrates the subjectivity–objectivity interface in the context of teaching science. Similarly, other topics of the science curriculum can also face similar dualities that are subject to refinement.

7.11 Open-Mindedness and not Relativity Helps in Understanding Objectivity

Objectivity and open-mindedness are indeed integral attributes of the scientific enterprise, but not in the sense generally presented in school science and textbooks. Objectivity consists not in denying preconceptions/presuppositions, but in the ability to modify beliefs in the light of emerging evidence and also encouraging open-mindedness. Scientists make errors and it is the community of scientists that helps to facilitate change by espousing open-mindedness. History of science shows that although scientists at a certain stage may have good reasons to believe that they need to go beyond objectivity, this does not represent relativity but rather open-mindedness. This serves to enhance the objectivity of collectively scrutinized scientific knowledge through decreasing the impact of individual scientists' idiosyncrasies and subjectivities. For example, although some creationists reject objectivity and relativize the truth of scientific knowledge, they are not necessarily open-minded. Another example of this aspect is the initial acceptance of the paramyxovirus as the causative agent of SARS and its replacement by the coronavirus, which illustrates not only the tentativeness of science but also skepticism and open-mindedness (Wong, Kwan, Hodson, & Jung, 2009). In the classroom, inclusion of such episodes from the history of science can facilitate a more meaningful pursuit of scientific inquiry.

7.12 Polanyi's Tacit Knowledge and Objectivity

According to Polanyi (1964, 1966), the rule bound knowing of empiricism and logic is linked to objectivity and the tacit knowing based on intuition and passion is linked to subjectivity. Consequently, personal knowledge is the unification of the objective and subjective aspects of scientific knowledge. In a similar vein, Daston and Galison (2007, p. 377) have endorsed Polanyi, by suggesting that logical positivism approximates to mechanical objectivity, whereas what scientists actually do (based on tacit knowledge) represents trained judgment. According to Guba and Lincoln (1989), tacit knowledge is all that we know minus all we can say, consequently, "... if the investigator is to be prohibited from using tacit knowledge (Polanyi, 1966) as he or she attempts to pry open this oyster of unknowns, the possibility of constructivist inquiry would be severely constrained, if not eliminated altogether" (p. 176). It is precisely the "tacit assumptions" that underlie the frameworks scientists use to design and develop their research programs that lead them to emphasize reason, empirical evidence, and objectivity. Furthermore, it seems that scientists are probably less reflective of "tacit assumptions" that guide their reasoning than most other intellectuals of the modern age (Blake, 1994). This shows that science education needs to recognize both mechanical objectivity and trained judgement and thus recognize the problematic nature of objectivity.

7.13 Positivism and Its Claims to Objectivity

School science generally promotes the idea that experiments provide data that reflected what was actually happening in the real world. Emphasizing such universal knowledge in the classroom based on a positivist perspective ignores the role of conflicting paradigms (controversies) and thus does not facilitate an understanding of how science progresses. Objectivist teaching strategies are heavily imbued with positivist epistemology that relies on algorithmic rather than conceptual understanding. In contrast, postpositivist perspectives in the philosophy of science (Phillips & Burbules, 2000) provide a better understanding by facilitating an integration of domain-specific information (plausibility of hypotheses) and domain-general aspects of the nature of science (heuristics that guide explorations). One possible sequence of a conceptual teaching strategy could be: setting up of a sequence, opening question, dialogue, conflicts (based on controversies), and negotiation of meaning. Objectivity, certainty, and infallibility as universal values of science may be challenged while studying controversies in their original historical context.

7.14 Reporting Style in Science as a False Guise of Objectivity

How we communicate science is an essential part of understanding "science in the making." Science and science education generally emphasize that researchers should maintain an objective voice (i.e., passive), and not to be passionately involved with their findings and interpretations. However, history of science shows that this is at best a chimera (Daston & Galison, 2007; Duhem, 1914; Hoffmann, 2012, 2014; Medawar, 1967), and research that matters is motivated by deep commitments and the passion to learn and understand. Indeed, reporting science involves a constant struggle between the theoretical frameworks of the scientist and the historian, as both are theory-laden. Given the influence exerted by editors and even the scientific community, scientists face a conflict with respect to using passive or active voice while reporting their findings. The active voice potentially recognizes the human dimension in data interpretation and knowledge construction. Given the complexity of the scientific enterprise, laboratory methods of gathering data and their interpretation may change over time due to some unforeseen findings. Still reporting of such research, written in retrospect is presented as highly consistent, rational, and logical from its inception. Consequently, reporting of a scientific event in a journal entails covering up the confusion, random, and chaotic means that produced it so as to give the impression that it represents an objective reflection of the world as it really exists. For science education the inclusion of the human element in the form of historical narratives is particularly helpful.

7.15 Role of Affect/Emotions and Objectivity

Studies of affect in science education are theoretically wide ranging and empirically diverse. In science education, emotions have generally been opposed by reason, truth, and the pursuit of objective knowledge. It is recommended that teachers (for that matter also students) should not express emotions as they are biased and thus there is no place for them in teaching and learning science. However, the difficulties involved in educational practice lead to satisfactions (when everything goes as planned) and frustrations (when things do not work as planned), and this necessarily leads to positive or negative emotions. Some emotions (such as happiness, pleasure, delight, thrill, and zeal) act to potentially enhance learning and optimize student achievement. Inclusion of affect and emotions in the classroom leads to an environment that is more in consonance with the history of science and the practice of science. The best solution to resolve this dilemma is perhaps through interactions among peers and also between the students and the teachers.

7.16   Scientific Method and Objectivity

The scientific method continues to be problematic in both science and science education, as it is generally believed that use of the scientific method ensures objectivity and the universality of science. To recognize that science is not culture-free is indeed a humbling experience for scientists. Gerald Holton (2014) recalling why he decided that Harvard Project Physics be based on a humanistic approach stated, "I based my decision in part on the hunch that more beginning students would come to take this course, to learn not only that F is equal to ma, but also that science is a fascinating part of human culture" (p. 1876). History of science shows that no set of objective rules or method can explain theory choice sufficiently. A scientist needs considerable experience to know under what circumstances and in what way any posited rules (formulated a priori) should be applied. Some science teachers believe that use of creativity and imagination (i.e., lack of a scientific method) during the interpretation phase of the data may compromise the objectivity of the scientists. On the contrary, history of science shows that it is precisely during the interpretation of the data that scientists need to be more creative. Lack of an understanding of the scientific enterprise (science in the making) that involves ambiguity, uncertainty, and intuitiveness (among other aspects) leads science educators and textbooks to emphasize the importance of the scientific method. Situating scientific inquiry in the context of "science in the making" leads to understanding complex and controversial subjects (such as evolutionary theory) more fruitful and even shows the problematic nature of progress in science. For example, students may think that Darwinism is not really a science at all but instead a worldview.

7.17   Social Interactions and the Evolving Nature of Objectivity

Social dimensions of science (e.g., peer-review process and interactions among members of the scientific community) facilitate the transition from a subjective to a more objective nature of scientific knowledge. Within this perspective, recognition of the social character of inquiry espouses pluralism, and acknowledges explanatory pluralism (Longino, 2002). Similarly, Giere (2006b) has recommended a pluralism of perspectives and that knowledge claims are perspectival rather than absolutely objective and hence cannot provide a "true" or "correct" answer to a problem. Pretensions of science to objectivity need to be countered with the social dimensions of knowledge. Errors in science are corrected by communication, first within the research group and later within the wider scientific community. In other words, objectivity in its purest sense is perhaps never an option, and is best understood within a social perspective based on sharing and communicating ideas. It is not the dualistic separation of objective and subjective knowledge (e.g., rational and creative, researcher, and researched) but rather the specific, social, cultural, and sociopolitical contexts that facilitate progress in science. Recent work on the life of Charles Darwin has shown that his theory of evolution was inextricably linked with its social dimensions. Knowledge is achieved primarily through a process of inquiry that is characterized by its social, experimental, and fallible nature. Nevertheless, it is not necessarily the experimental data (Baconian orgy of quantification) but rather the diversity/plurality of ideas in a scientific discipline that contributes toward a better understanding of the evolving nature of objectivity. In essence, the pluralist approach dissolves the distinction between the epistemic and the social (Longino, 1990) and thus helps to correct flaws and enhance the reliability of scientific results. Although within Marxism the influence of social factors is important, instead in Mainland China Mao's concept of "practice" is highly valued. The role of social factors and the scientific community is important and at times objectivity may become synonymous with consensus. History of science, however, shows that there is no guarantee that the scientific community is infallible (cf. Rowlands, Graham, & Berry, 2011). All knowledge develops and forms part of the social, cultural, and local milieu. Given appropriate social interactions, the idea of localness can transcend and facilitate trans-localness, which leads to greater objectivity. In this context, Daston and Galison's (2007) concept of the evolving nature of objectivity, which facilitates the different forms of objectivity (scientific judgment) to coexist and even perhaps compete.

7.18 Theory-Laden Observations and Objectivity

The role of theory-laden observations is important as school science fosters the idea that experimental observations are entirely objective. History of science provides many intriguing episodes. For example, in the 1919 solar eclipse expedition, if Edington had not been aware of Einstein's special theory of relativity, it would have been extremely difficult to interpret the observations. In this context data obtained by students in an experiment can provide grounds for relating the experimental observations and students' prior beliefs. Experiments are difficult to conduct and can provide evidence for more than one hypothesis, and students are generally unaware of this possibility.

7.19 Transgression, Objectivity and Scientific Progress at a Crossroads

This section is primarily based on results reported in Chap. 6 (based on general chemistry textbook evaluations) and following are some of the salient features:

(a) Due to the controversies and interactions among members of the scientific community, objectivity itself is achieved partially, progressively, in degrees, and hence the need for "transgression of objectivity."
(b) If objectivity is achieved in degrees it is plausible to suggest that the scientific method based on a series of rigid steps cannot characterize the scientific endeavor.

(c) As the word "method" itself denotes a more structured approach to science, it is preferable instead to emphasize the work of scientists themselves within a historical context.

(d) It is important to note that even before the Scanning tunneling microscopy (STM) was invented, scientists (e.g., Dalton and many others) were trying to understand atomic structure through indirect experiments (cf. Hoffmann, 2012). This shows that the quest for knowledge/understanding of matter has a long history, starting perhaps with magnifying glasses in the fifteenth century.

(e) Presentations of some textbooks give the impression that STM provides actual photographs of the atoms, whereas in actual practice the images are computer generated.

(f) STM and Atomic force microscopy (AFM) investigate only surface atoms and do not provide information with respect to internal structure of atoms.

(g) STM can be used only for conductive surfaces, whereas AFM can be used with almost any surface.

(h) Some textbooks raised the question: are atoms real? And that we can now "see" atoms and also their magnifications (up to 10 million times). It is plausible to suggest that if atoms are real that refers to "representation," and "seeing" and the magnifications to "presentation"—thus scientific progress is at a crossroads. In other words, the balance has shifted toward presentation that facilitates intervention (nanotechnology).

(i) Some textbooks emphasized the production of new materials based on nanotechnology that were previously even difficult to dream of, such as: C_{60}, buckminsterfullerene, the third form of carbon; artificial cells that can provide additional oxygen to the bloodstream; miniaturizing of electrical instruments (cell phones, computers); waterproof and wrinkle-free nanoparticle based textile products; nanoscale materials that are highly conductive (e.g., gold, which is otherwise not a conductor) and some even 40 times stronger than steel; hydrogen absorbed into nanotubes would solve the problems associated with hydrogen fuel cells; and enormous surface to volume ratio of nanomaterials is of special importance for the drug industry.

These innovations in nanotechnology provide examples of cutting-edge research that is at a crossroads with our existing state of knowledge, and even perhaps seem to belong to the realm of science fiction. However, as suggested by Hoffmann (Email to author, February 24, 2016b) a word of caution is necessary in understanding the significance and future prospects of nanotechnology.

7.20 Uncertainty and Objectivity

In classroom practice positivism imbues scientific knowledge with a Laplacian certainty denied to all other disciplines. This leads to teaching science by neglecting the social and cultural milieu in which scientists work and the certainty surrounding science is conveyed as a dogma. According to *Project* 2061: "The notion that scientific knowledge is always subject to modification can be difficult for students to grasp. It seems to oppose the certainty and truth popularly accorded to science, and runs counter to the yearning for certainty that is characteristic of most cultures, perhaps especially so among youth" (AAAS, 1993, p. 5). Actually, in students' processes of construction of knowledge uncertainty can help to advance the learning process. The knowledge that has already been acquired allows the researchers to raise new questions because there is uncertainty in existing knowledge. The dynamics of uncertainty and raising new questions helps to facilitate greater understanding. Based on Piaget's genetic epistemology, constructivism emphasizes the inherent uncertainty of the constructed knowledge of the world by both children and scientists. Furthermore, the concept of "objectivity" is reconceptualized as consensual agreement among scientific communities of practices, quite similar to what Daston and Galison (2007) have referred to as "trained judgment."

7.21   Is Objectivity an Opiate of the Academic?

In the light of the results presented in this book, it is important to consider the following thesis put forward by Aikenhead (2008): "Given the prominence of the objectivity/subjectivity dichotomy in science education and most of its research, many academics must feel comfortable with the dichotomy, so much so that I wonder if objectivity has become the opiate of the academic" (p. 584). This is a controversial thesis and perhaps many science educators may consider it to be too radical and extreme. Nevertheless, let us reconsider the results reported in this book in order to have a better perspective. Chap. 6 showed that almost 90% of general chemistry textbook authors (published in USA) did not recognize the problematic nature of objectivity and again about half endorsed the traditional step-wise scientific method. About one-third of the authors of articles written by science education researchers (Chaps. 3–5) did not recognize the problematic nature of objectivity. Given this state of affairs and perhaps with some reluctance, I would like to endorse Aikenhead's (2008) thesis, namely "objectivity as an opiate of the academic." Lest it be misconstrued, my objective (as part of the science education community) in raising this issue is that of a constructive criticism and a call for a critical appraisal of how we do and understand science, while trying to grasp the evolving nature of objectivity. At this stage it is important to note that after reading a preliminary version of this chapter, Aikenhead (personal communication, July 27, 2016) suggested the following: "It would be interesting to read a parallel chapter to your Chap. 7 from the standpoint of subjectivity. On pages 9–10 you explore the two poles of a continuum. In 1991, I published a grade 10 STS science textbook *Logical Reasoning in Science and Technology*, in which objectivity was replaced by the notion of degrees of subjectivity. Thus, the value that guides scientists is to reach the lowest level of subjectivity as humanly and financially possible. This stance makes intuitive sense to high school students, and it eliminates many of the issues that arise in Chap. 7 about the problems with objectivity" (underline added). The notion of "degrees of subjectivity" can be compared to what Machamer and Wolters (2004, pp. 9–10) have referred to as "objectivity comes in degrees." In a similar vein Daston and Galison (2007, p. 374) have pointed out that, "subjectivity is not a weakness of the self, [but rather] it is the self." For a science educator it is important to understand that the notions of subjectivity and objectivity are intricately intertwined and it is the constant struggle between these two poles of a continuum that facilitates progress in science.

Finally, it is concluded that the evolving nature of objectivity is important for science education as school and college science generally simplify complex historical episodes under the rubric of objectivity without really understanding that the underlying issues perhaps are dependent on trained judgment (Daston & Galison, 2007). Although, achievement of objectivity in actual scientific practice is a myth, it still remains a powerful and useful idea (Harding, 2015). Similarly, Aikenhead and Michell (2011) have endorsed a similar approach: "Consensus making *reduces* the subjectivity of individual scientists or teams of scientists. Consequently, a realistic goal for scientists is *low subjectivity*. The public storyline that scientists attain objectivity is a myth .... The ideal of objectivity fails in the reality of practice .... Nevertheless, objectivity remains a powerful and useful ideal" (p. 41, underline added). Despite a critical stance toward the role played by objectivity, it is important to note that scholars of different disciplines and persuasion still consider it to be a useful epistemic virtue (e.g., Aikenhead, Daston, Galison, Harding, Hodson, Hoffmann, & Machamer). It is essential that science educators debated these epistemic virtues in order to clarify what they entail and thus facilitate a better understanding of the dynamics of scientific progress.

7.22   Educational Implications

Based on different chapters of this book, here I summarize the following educational implications that can facilitate the work of both students and teachers:

- Studying controversies in their original scientific/historical context of inquiry can facilitate a perspective that can help to question objectivity in different topics of the science curriculum.
- Differentiating between scientists' theory and historians' theory can provide a better understanding of how the evolving nature of objectivity is crucial for following scientific progress.
- Despite the importance of experimental data scientists can still interpret the same data differently. This shows that experimental data do not necessarily facilitate objectivity in science.
- Experimental facts remain mute unless an attempt is made to interpret them, which leads to the elaboration of a narrative that facilitates understanding.
- Pluralism of perspectives (Giere, Longino) helps to correct flaws and thus enhance the reliability of scientific results. Pluralism based on value-judgments

is a virtue rather than a liability. Recognizing and evaluating value-laden science is important for understanding progress.

- The need to understand objectivity more as a process rather than an end product. For Piaget a process consists of successive approximations toward objectivity, and furthermore objectivity comes in degrees (Daston, Galison, Machamer, & Wolters).
- Articulation of tacit knowledge (Polanyi) in contrast to rigid rules and algorithms is more helpful in understanding objectivity.
- Differentiation between algorithmic and conceptual teaching strategies. Algorithmic strategies are based on adherence to rigid rules and procedures that approximate to mechanical objectivity. In contrast, conceptual strategies can generate cognitive conflicts and thus are open to negotiation of meaning that is trained judgment.
- Tentative nature of science is an important characteristic of nature of science. For example, atomic models have changed over the last 200 years (Dalton, Thomson, Rutherford, Bohr, Sommerfeld, wave mechanical). It is plausible to suggest that the tentativeness of science manifests itself in the evolving nature of objectivity. For example, at some stage in history all atomic models were considered to be objective, especially to its proponents. Teaching tentativeness of science in the context of the evolving nature of objectivity can facilitate a better understanding of scientific progress.
- School science generally associates and emphasizes certainty and objectivity with progress in science. However, it can be argued that lack of certainty can be used as a means to facilitate conceptual understanding. Acquired knowledge raises further questions that need research and hence show uncertainty, which can drive the learning process of acquiring knowledge.
- Given the evolving nature of objectivity it is important that teachers consider themselves also as learners and that their constructions (classroom interventions) of knowledge are never complete but rather tentative.
- In the history of science one form of objectivity did not supplant the other, but rather the two coexisted. Consequently, it is plausible to suggest that classroom discussions could provide an opportunity to facilitate and understand the objectivity–subjectivity continuum.

References

Aikenhead, G. (2008). Objectivity: the opiate of the academic? *Cultural Studies of Science Education, 3*(3), 581–585.

Aikenhead, G., & Michell, H. (2011). *Bridging cultures:* indigenous and scientific ways of knowing nature. Toronto: Pearson Education Canada.

American Association for the Advancement of Science, AAAS. (1993). *Benchmarks for science literacy: project 2061.* Washington: Oxford University Press.

Blake, D. D. (1994). Revolution, revision or reversal: genetics-ethics curriculum. *Science & Education, 3*(4), 373–391.

Cushing, J. T. (1995). Hermeneutics, underdetermination and quantum mechanics. *Science & Education, 4*(2), 137–147.

Daston, L., & Galison, P.L. (1992). *The image of objectivity. Representations, 40* (special issue: seeing science), 81–128.

Daston, L., & Galison, P. (2007). *Objectivity*. New York: Zone Books.

Duhem, P. (1914). *The aim and structure of physical theory* (2nd ed., trans: Wiener, P. P.). New York: Atheneum.

Gergen, K. J. (1994). The mechanical self and the rhetoric of objectivity. In A. Megill (Ed.), *Rethinking objectivity*. Durham: Duke University Press.

Giere, R. N. (2006a). Perspectival pluralism. In S. H. Kellert, H. E. Longino & C. K. Waters (Eds.), *Scientific pluralism* (pp. 26–41). Minneapolis: University of Minnesota Press.

Giere, R. N. (2006b). *Scientific perspectivism*. Chicago: University of Chicago Press.

Guba, E. G., & Lincoln, Y. S. (1989). Fourth generation evaluation. Newbury Park: Sage.

Harding, S. (2015). *Objectivity and diversity: another logic of scientific research*. Chicago: University of Chicago Press.

Hoffmann, R. (2012). J. Kovac & M. Weisberg (Eds.), *Roald Hoffmann on the philosophy, art, and science of chemistry*. New York: Oxford University Press.

Hoffmann, R. (2014). The tensions of scientific storytelling: science depends on compelling narratives. *American Scientist, 102*, 250–253.

Holton, G. (1978a). Subelectrons, presuppositions, and the Millikan-Ehrenhaft dispute. *Historical Studies in the Physical Sciences, 9*, 161–224.

Holton, G. (1978b). *The scientific imagination: case studies*. Cambridge: Cambridge University Press.

Holton, G. (2014). The neglected mandate: teaching science as part of our culture. *Science & Education, 23*, 1875–1877.

Irzik, G. (2015). Values and Western science knowledge. In R. Gunstone (Ed.), *Encyclopedia of science education* (pp. 1093–1096). Heidelberg: Springer.

Longino, H. E. (1990). *Science as social knowledge: values and objectivity in scientific inquiry*. Princeton: Princeton University Press.

Longino, H. E. (2002). *The fate of knowledge*. Princeton: Princeton University Press.

Machamer, P., & Wolters, G. (2004). Introduction: science, values and objectivity. In P. Machamer & G. Wolters (Eds.), *Science, values and objectivity* (pp. 1–13). Pittsburgh: University of Pittsburgh Press.

McCarthy, C.L. (2014). Cultural studies in science education: philosophical considerations. In M.R. Matthews (Ed.), *International handbook of research in history, philosophy and science teaching* (Vol. III, pp. 1927–1964).

Medawar, P. B. (1967). *The art of the soluble*. London: Methuen.

Merton, R.K. (1942). Science and technology in a democratic order. *Journal of Legal and Political Sociology, 1*. Reprinted as 'Science and Democratic Social Structure', in his *Social theory and social structure*. New York: Free Press (1957).

Niaz, M. (2016). *Chemistry education and contributions from history and philosophy of science*. Dordrecht: Springer.

Phillips, D. C., & Burbules, N. C. (2000). *Postpositivism and educational research*. New York: Rowman & Littlefield.

Piaget, J. (1971). *Biology and knowledge: an essay on the relations between organic regulations and cognitive processes*. Chicago: University of Chicago Press.

Polanyi, M. (1964). *Personal knowledge: towards a post-critical philosophy*. Chicago: University of Chicago Press. (first published 1958).

Polanyi, M. (1966). *The tacit dimension*. London: Routledge & Kegan Paul.

Reiss, M. J. (2014). What significance does Christianity have for science education? In M. R. Matthews (Ed.), *International handbook of research in history, philosophy and science teaching* (pp. 1637–1662). Dordrecht: Springer.

Rowlands, S., Graham, T., & Berry, J. (2011). Problems with fallibilism as a philosophy of mathematics education. *Science & Education, 20*(7–8), 625–654.

Slezak, P. (1994). Sociology of scientific knowledge and scientific education, Part I. *Science & Education, 3*(3), 265–294.

Smith, M. U., & Siegel, H. (2016). On the relationship between belief and acceptance of evolution as goals of evolution education. *Science & Education, 25*(5–6), 473–496.

Vermeir, K. (2013). Scientific research: commodities or commons? *Science & Education, 22*(10), 2485–2510.

Wong, S. L., Kwan, J., Hodson, D., & Jung, B. H. W. (2009). Turning crisis into opportunity: nature of science and scientific inquiry as illustrated in the scientific research on severe acute respiratory syndrome. *Science & Education, 18*(1), 95–118.

Appendix 1

Articles from the journal *Science & Education* (Springer) evaluated in this study

Abd-El-Khalick, F. (2013). Teaching *with* and *about* nature of science, and science teacher knowledge domains. *Science & Education, 22*(9), 2087–2107.

Allchin, D. (1999). Values in science: An educational perspective. *Science & Education, 8*(1), 1–12.

Allchin, D. (2004). Pseudohistory and pseudoscience. *Science & Education, 13*(3), 179–195.

Allgaier, J. (2010). Scientific experts and the controversy about teaching creation/evolution in the UK press. *Science & Education, 19*(6-8). 797–819.

Blake, D.D. (1994). Revolution, revision or reversal: Genetics-ethics curriculum. *Science & Education, 3*(4), 373–391.

Blanco, M.P. (2014). "Palabras de la ciencia": Pedro Castera and scientific writing in Mexico's *fin the siècle. Science & Education, 23*(3), 541-556.

Carolino, L.M. (2012). Measuring the heavens to rule the territory: Felipe Folque and the teaching of astronomy at the Lisbon Polytechnic school and the modernization of the state apparatus in nineteenth century Portugal. *Science & Education, 21*(1), 109-133.

Carrier, M. (2013). Values and objectivity in science: Value-ladenness, pluralism and the epistemic attitude. *Science & Education, 22*(10), 2547-2568.

Cartwright, J. (2007). Science and literature: Towards a conceptual framework. *Science & Education, 16*(2), 115–139.

Chamizo, J.A. (2013). A new definition of models and modeling in chemistry's teaching. *Science & Education, 22*(7), 1613–1632.

Cobern, W.W. (1995). Science education as an exercise in foreign affairs. *Science & Education, 4*(3), 287–302.

Cobern, W.W., & Loving, C.C. (2008). An essay for educators: Epistemological realism really is common sense. *Science & Education, 17*(4), 425–447.

Cordero, A. (1992). Science, objectivity and moral values. *Science & Education, 1*(1), 49–70.

Cordero, A. (2012). Mario Bunge's scientific realism. *Science & Education, 21*(10), 1419–1436.

Crasnow, S. (2008). Feminist philosophy of science: "Standpoint" and knowledge. *Science & Education, 17*(10), 1089–1110.

Cushing, J.T. (1995). Hermeneutics, underdetermination and quantum mechanics. *Science & Education, 4*(2), 137–147.

© Springer International Publishing AG 2018

M. Niaz, *Evolving Nature of Objectivity in the History of Science and its Implications for Science Education*, Contemporary Trends and Issues in Science Education,

DOI 10.1007/978-3-319-67726-2

Dahlin, B. (2001). The primacy of cognition — or of perception? A phenomenological critique of the theoretical bases of science education. *Science & Education, 10*(5), 453–475.

Davson-Galle, P. (2002). Science, values and objectivity. *Science & Education, 11*(2), 191–202.

Deng, F., Chai, C.S., Tsai, C.-C., & Lin, T.-J. (2014). Assessing South China (Guangzhou) high school students' views on nature of science: A validation study. *Science & Education, 23*(4), 843–863.

Depew, D.J. (2010). Darwinian controversies: An historiographical recounting. *Science & Education, 19*(4–5), 323–366.

Develaki, M. (2007). The model-based view of scientific theories and the structuring of school science programmes. *Science & Education, 16*(7–8), 725–749.

Develaki, M. (2008). Social and ethical dimension of the natural sciences, complex problems of the age, interdisciplinarity, and the contribution of education. *Science & Education, 17*(8–9), 873–888.

Develaki, M. (2012). Integrating scientific methods and knowledge into the teaching of Newton's theory of gravitation: An instructional sequence for teachers' and students' nature of science education. *Science & Education, 21*(6), 853–879.

Eger, M. (1993). Hermeneutics as an approach to science: Part II. *Science & Education, 2*(4), 303–328.

El-Hani, C.N. (2015). Mendel in genetics teaching: Some contributions from history of science and articles for teachers. *Science & Education, 24*(1–2), 173–204.

Erduran, S., & Mugaloglu, E.Z. (2013). Interactions of economics of science and science education: Investigating the implications for science teaching and learning. *Science & Education, 22*(10), 2405–2425.

Ernest, P. (1992). The nature of mathematics: Towards a social constructivist account. *Science & Education, 1*(1), 89–100.

Fiss, A. (2012). Problems of abstraction: Defining an American standard for mathematics education at the turn of the twentieth century. *Science & Education, 21*(8), 1185–1197.

Ford, M. (2008). 'Grasp of practice' as a reasoning resource for inquiry and nature of science understanding. *Science & Education, 17*(2–3), 147–177.

Galili, I. (2011). Promotion of cultural content knowledge through the use of the history and philosophy of science. *Science & Education, 21*(9), 1283–1316.

Garrison, J. (1997). An alternative to Von Glasersfeld's subjectivism in science education: Deweyan social constructivism. *Science & Education, 6*(3), 301–312.

Garrison, J. (2000). A reply to Davson-Galle. *Science & Education, 9*(6), 615–620.

Gauch, H.G. (2009). Science, worldviews and education. *Science & Education, 18*(6-7), 667–695.

Gauld, C.F. (2005). Habits of mind, scholarship and decision making in science and religion. *Science & Education, 14*(3–5), 291–308.

Gil-Pérez, D., Vilches, A., Fernández, I., Cachapuz, A., Praia, J., Valdés, P., Salinas, J. (2005). Technology as 'applied science': A serious misconception that reinforces distorted and impoverished views of science. *Science & Education, 14*(3-5), 309–320.

Ginev, D.J. (2008). Hermeneutics of science and multi-gendered science education. *Science & Education, 17*(10), 1139–1156.

Hadzigeorgiou, Y., & Schulz, R. (2014). Romanticism and romantic science: Their contribution to science education. *Science & Education, 23*(10), 1963–2006.

Hadzidaki, P. (2008a). 'Quantum mechanics' and 'scientific explanation' an explanatory strategy aiming at providing 'understanding.' *Science & Education, 17*(1), 49–73.

Heffron, J.M. (1995). The knowledge most worth having: Otis W. Caldwell (1869–1947) and the rise of the general science course. *Science & Education, 4*(3), 227–252.

Hildebrand, D., Bilica, K., & Capps, J. (2008). Addressing controversies in science education: A pragmatic approach to evolution education. *Science & Education, 17*(8–9), 1033–1052.

Homchick, J. (2010). Objects and objectivity: The evolution controversy at the American museum of natural history, 1915-1928. *Science & Education, 19*(4–5), 485–503.

Howard, D. (2009). Better red than dead — Putting an end to the social irrelevance of postwar philosophy of science. *Science & Education, 18*(2), 199–220.

Intemann, K. (2008). Increasing the number of feminist scientists: Why feminist aims are not served by the underdetermination thesis. *Science & Education, 17*(10), 1065–1079.

Irzik, G., & Nola, R. (2011). A family resemblance approach to the nature of science for science education. *Science & Education, 20*(7–8), 591–607.

Jiménez-Aleixandre, M.P. (2014). Determinism and underdetermination in genetics: Implications for students' engagement in argumentation and epistemic practices. *Science & Education, 23*(2), 465–484.

Kipnis, N. (2007). Discovery in science and teaching science. *Science & Education, 16*(9–10), 883–920.

Kitchener, R.F. (1993). Piaget's epistemic subject and science education: Epistemological versus psychological issues. *Science & Education, 2*(2), 137–148.

Kolstø, S.D. (2008). Science education for democratic citizenship through the use of the history of science. *Science & Education, 17*(8–9), 977–997.

Kosso, P. (2009). The large-scale structure of scientific method. *Science & Education, 18*(1), 33–42.

Krogh, L.B., & Nielsen, K. (2013). Introduction: How science works — and how to teach it. *Science & Education, 22*(9), 2055–2065.

Lau, K.-C., & Chan, S.-L. (2013). Teaching about theory-laden observation to secondary students through manipulated lab inquiry experience. *Science & Education, 22*(10), 2641–2658.

Legates, D.R., Soon, W., Briggs, W.M., Monckton of Brenchley, C. (2015). Climate consensus and 'misinformation': A rejoinder to *Agnotology, scientific consensus, and the teaching and learning of climate change*. *Science & Education*, in press.

Leite, L. (2002). History of science in science education: Development and validation of a checklist for analyzing the historical content of science textbooks. *Science & Education, 11*(4), 333–359.

Lindahl, M.G. (2010). Of pigs and men: Understanding students' reasoning about the use of pigs as donors for xenotransplantation. *Science & Education, 19*(9), 867–894.

Lövheim, D. (2014). Scientists, engineers and the society of free choice: Enrollment as policy and practice in Swedish science and technology education 1960–1990. *Science & Education, 23*(9), 1763–1784.

Lyons, S.L. (2010). Evolution and education: Lessons from Thomas Huxley. *Science & Education, 19*(4–5), 445–459.

Machamer, P, & Woody, A. (1994). A model of intelligibility in science: Using Galileo's balance as a model for understanding the motion of bodies. *Science & Education, 3*(3), 215–244.

Marroum, R.-M. (2004). The role of insight in science education: An introduction to the cognitional theory of Bernard Lonegran. *Science & Education, 13*(6), 519–540.

Matthews, M.R. (1992). History, philosophy, and science teaching: The present reapproachment. *Science & Education, 1*(1), 11–47.

Nielsen, K.H. (2013). Scientific communication and the nature of science. *Science & Education, 22*(9), 2067–2086.

Park, H., Nielsen, W., & Woodruff, E. (2014). Students' conceptions of the nature of science: Perspectives from Canadian and Korean middle school students. *Science & Education, 23*(5), 1169–1196.

Patronis, T., & Spanos, D. (2013). Exemplarity in mathematics education: From a romanticist viewpoint to a modern hermeneutical one. *Science & Education, 22*(8), 1993–2005.

Pennock, R.T. (2002). Should creationism be taught in the public schools? *Science & Education, 11*(2), 111–133.

Pennock, R.T. (2010). The postmodern sin of intelligent design creationism. *Science & Education, 19*(6–8).

Pospiech, G. (2003). Philosophy and quantum mechanics in science teaching. *Science & Education, 12*(5–6), 559–571.

Quílez, J. (2009). From chemical forces to chemical rates: A historical/philosophical foundation for the teaching of chemical equilibrium. *Science & Education, 18*(9), 1203–1251.

Rowell, J.A. (1993). Developmentally-based insights for science teaching. *Science & Education, 2*(2), 111–136.

Rowlands, S., Graham, T., & Berry, J. (2011). Problems with fallibilism as a philosophy of mathematics education. *Science & Education, 20*(7–8), 625–654.

Russanen, A.-M., Pöyhönen, S. (2013). Concepts in change. *Science & Education, 22*(6), 1389–1403.

Sievers, K.H. (1999). Toward a direct realist account of observation. *Science & Education, 8*(4), 387–393.

Skordoulis, C.D. (2008). Science and worldviews in the Marxist tradition. *Science & Education, 17*(6), 559–571.

Silverman, M.P. (1992). Raising questions: Philosophical significance of controversy in science. *Science & Education, 1*(2), 163–179.

Slezak, P. (1994). Sociology of scientific knowledge and scientific education, Part I. *Science & Education, 3*(3), 265–294.

Smith, M.U., Siegel, H., & McInerney, J.D. (1995). Foundational issues in evolution education. *Science & Education, 4*(1), 23–46.

Suchting, W.A. (1992). Constructivism deconstructed. *Science & Education, 1*(3), 223–254.

Takacs, P., & Ruse, M. (2013). The current status of the philosophy of biology. *Science & Education, 22*(1), 5–48.

Talanquer, V. (2013). School chemistry: The need for transgression. *Science & Education, 22*(7), 1757–1773.

Uebel, T.E. (2004). Education, enlightenment and positivism: The Vienna Circle's scientific world-conception revisited. *Science & Education, 13*(1–2), 41–66.

Vermeir, K. (2013). Scientific research: Commodities or commons? *Science & Education, 22*(10), 2485–2510.

Wan, Z.H., Wong, S.L., & Zhan, Y. (2013). When nature of science meets Marxism: Aspects of nature of science taught by Chinese science teacher educators to prospective science teachers. *Science & Education, 22*(5), 1115–1140.

Wong, S.L., Kwan, J., Hodson, D., & Jung, B.H.W. (2009). Turning crisis into opportunity: Nature of science and scientific inquiry as illustrated in the scientific research on severe acute respiratory syndrome. *Science & Education, 18*(1), 95–118.

Appendix 2

Distribution of articles (*Science & Education*) according to author's area of research, context of the study and level (classification)

No.	Authors in the reference	Author's area of research	Context of the study	Level
1	Abd-El-Khalick, F. (2013)	Science education	Nature of science and teacher knowledge	IV
2	Allchin, D. (1999)	Philosophy of science	Values in science	III
3	Allchin, D. (2004)	Philosophy of science	Craniology & phrenology as pseudoscience	III
4	Allgaier, J. (2010)	Sociology of science	Creation-evolution controversy	III
5	Blake, D.D. (1994)	Biology education	Science & ethics in genetics education	III
6	Blanco, M.P. (2014)	Literature	Science fiction	II
7	Carolino, L.M. (2012)	History of science	Teaching astronomy	II
8	Carrier, M. (2013)	Philosophy of science	Values, pluralism and objectivity	V
9	Cartwright, J. (2007)	Biology education	Literature and science	II
10	Chamizo, J.A. (2013)	Chemistry education	Models in chemistry teaching	II
11	Cobern, W.W. (1995)	Science education	Social/cultural milieu	III
12	Cobern, W.W., & Loving, C.C. (2008)	Science education	Epistemological realism	III
13	Cordero, A. (1992)	Philosophy of science	Philosophy of science	III
14	Cordero, A. (2012)	Philosophy of science	Bunge's scientific realism	III

(continued)

(continued)

No.	Authors in the reference	Author's area of research	Context of the study	Level
15	Crasnow, S. (2008)	Philosophy of science	Feminist philosophy of science	III
16	Cushing, J.T. (1995)	Philosophy of physics	Contingency and quantum mechanics	III
17	Dahlin, B. (2001)	Science education	Phenomenology and science education	II
18	Davson-Galle, P. (2002)	Science education	Values and objectivity	II
19	Deng, F., Chai, C.S., Tsai, C.-C., & Lin, T.-J. (2014)	Science education	NOS views of Chinese students	II
20	Depew, D.J. (2010)	Philosophy of science	Darwinian controversies	III
21	Develaki, M. (2007)	Science education	Model-based view of scientific theories	II
22	Develaki, M. (2008)	Science education	Social & ethical dimensions of science	III
23	Develaki, M. (2012)	Science education	Newton's theory of gravitation	III
24	Eger, M. (1993)	Physics	Hermeneutics	II
25	El-Hani, C.N. (2015)	Biology education	Mendel in genetics teaching	II
26	Erduran, S., & Mugaloglu, E.Z. (2013)	Chemistry education	Economics of science & science education	II
27	Ernest, P. (1992)	Mathematics education	Social constructivism & mathematics	II
28	Fiss, A. (2012)	Science studies	Mathematics education & history of science	III
29	Ford, M. (2008)	Science education	Understanding NOS	III
30	Freire, O. (2003)	History of physics	Controversy in quantum physics	II
31	Galili, I. (2011)	Physics education	Cultural context of knowledge	IV
32	Galili, I. (2013)	Physics education	Imagery in science education	II
33	Garrison, J. (1997)	Educational philosophy	Deweyan social constructivism	II
34	Garrison, J. (2000)	Educational philosophy	Constructivism	II
35	Gauch, H.G. (2009)	Philosophy of science	Science & worldviews	III
36	Gauld, C.F. (2005)	Education	Science & religion	II
37	Gil-Pérez, D., Vilches, A., Fernández, I., Cachapuz, A., Praia, J., Valdés, P., & Salinas, J. (2005)	Science education	Science-technology relationship	III

(continued)

(continued)

No.	Authors in the reference	Author's area of research	Context of the study	Level
38	Ginev, D. J. (1995)	Philosophy	Hermeneutic conception of science	II
39	Ginev, D.J. (2008)	Philosophy	Multi-gendered science	III
40	Glasersfeld, E.V. (1992)	Psychology	Constructivism	II
41	Good, R., & Shymansky, J. (2001)	Science education	Science literacy: relativist or realist	II
42	Goodney, D.E., & Long, C.S. (2003)	Chemistry	Scientific literacy based on historical texts	III
43	Grandy, R., & Duschl, R. A. (2007)	Philosophy of science	Inquiry in school science	III
44	Hadzidaki, P. (2008a)	Science education	Understanding quantum mechanics	III
45	Hadzidaki, P. (2008b)	Science education	Heisenberg microscope & NOS	III
46	Hadzigeorgiou, Y. (2015)	Science	Science education as socio-political action	II
47	Hadzigeorgiou, Y., & Schulz, R. M. (2014)	Science education	Romanticism and science education	III
48	Heelan, P.A. (1995)	Philosophy	Quantum mechanics and hermeneutics	II
49	Heffron, J.M. (1995)	History of education	General science courses & science education	III
50	Hildebrand, D., Bilica, K., & Capps, J. (2008)	Philosophy	Controversy in science education	III
51	Hoffman, M. (2013)	Science education	General science courses	II
52	Homchick, J. (2010)	Writing & rhetoric	Evolutionary theory	V
53	Howard, D. (2009)	Philosophy	Social nature of scientific knowledge	III
54	Intemann, K. (2008)	History & philosophy	Feminist values & under determination	III
55	Irzik, G. (2013)	Philosophy of science	Commercialization of science	II
56	Irzik, G., & Nola, R. (2011)	Philosophy of science	Family resemblance & nature of science	III
57	Jiang, F., & McComas, W.F. (2014)	Science education	Nature of science in popular science books	III
58	Jiménez-Aleixandre, M.P. (2014)	Biology education	Argumentation in genetics	IV
59	Jorgensen, L.M., & Ryan, S.A. (2004)	Science education	Relativism, values & morals	II

(continued)

(continued)

No.	Authors in the reference	Author's area of research	Context of the study	Level
60	Jung, W. (2012)	Physics education	Philosophy of science & education	II
61	Kendig, C. (2013)	Philosophy	Integrating history & philosophy of science	II
62	Kipnis, N. (2007)	History of science	Discovery in science	II
63	Kirschner, P.A. (1992)	Science education	Practical work in science	III
64	Kitchener, R.F. (1993)	Philosophy of science	Piaget's epistemic subject	II
65	Kolstø, S.D. (2008)	Physics education	Science education & democratic citizenship	II
66	Kosso, P. (2009)	Philosophy	Scientific method	I
67	Krogh, L.B., & Nielsen, K. (2013)	Science education	Functional scientific literacy	III
68	Kruckeberg, R. (2006)	Science education	Constructivism & Dewey	I
69	Kubli, F. (2007)	Physics education	Experiments and stories in science	III
70	Lacey, H. (2009)	Philosophy of science	World views & values	II
71	Lau, K.-C., & Chan, S.-L. (2013)	Science education	Teaching theory-laden observation	IV
72	Lawson, A.E. (2000)	Biology education	Nature of knowledge	II
73	Legates, D.R., Soon, W., Briggs, W.M., Monckton of Brenchley, C. (2015)	Geography	Consensus & climate change	III
74	Leite, L. (2002)	Science education	History of science & textbooks	III
75	Levinson, R. (2008)	Science education	Socio-scientific issues	III
76	Levrini, O., Bertozzi, E., Gagliardi, M., Tomasini, N.G., Pecovi, B., Tasquier, G., & Galili, I. (2014)	Physics education	Discipline-culture framework	II
77	Lindahl, M.G. (2009)	Science education	Ethics & morals	II
78	Lindahl, M.G. (2010)	Science education	Expert knowledge	II
79	Lövheim, D. (2014)	Education	Enrollment practice	II
80	Lyons, S.L. (2010)	History of science	Evolution & education	III
81	Machamer, P., & Woody, A. (1994)	Philosophy of science	Model of intelligibility	III
82	Marroum, R.-M. (2004)	Physics education	Insight in science education	II
83	Matthews, M.R. (1992)	Science education	History & objectivity	IV

(continued)

(continued)

No.	Authors in the reference	Author's area of research	Context of the study	Level
84	Matthews, M.R. (2004)	Science education	Reappraising positivism	II
85	McComas, W.F. (2008)	Biology education	Historical examples & nature of science	II
86	Metz, D., Klassen, S., McMillan, B., & Clough, M. (2007)	Science education	Historical narratives	III
87	Mugaloglu, E.Z. (2014)	Science education	Pseudo-science & constructivism	II
88	Niaz, M. (2009)	Science education	Nature of science based on historical controversies	III
89	Nielsen, K.H. (2013)	Science studies	Science in the making	III
90	Oliveira, M.B. (2013)	Science & technology	Commodification of science	II
91	Park, H., Nielsen, W., & Woodruff, E. (2014)	Science education	Nature of science	III
92	Patronis, T., & Spanos, D. (2013)	Mathematics education	Hermeneutics	III
93	Pauri, M. (2003)	Philosophy of physics	Quantum theory	II
94	Pennock, R.T. (2002)	Philosophy	Creationism & school science	II
95	Pennock, R.T. (2010)	Philosophy	Postmodernism & intelligent design	III
96	Phillips, D.C. (2004)	Philosophy of science	Positivism & science education	III
97	Pinnick, C. (2008)	Philosophy of science	Feminist theory	II
98	Pospiech, G. (2003)	Physics education	Quantum mechanics & philosophy	III
99	Quale, A. (2002)	Science education	Metaphors & constructivism	II
100	Quale, A. (2007)	Science education	Radical constructivism & relativism	II
101	Quílez, J. (2009)	Chemistry education	Chemical equilibrium & historical context	III
102	Reisch, G. (2009)	Philosophy of science	Political engagement & philosophy of science	I
103	Roscoe, K. (2004)	Education	Constructivism	II
104	Rowell, J.A. (1993)	Science	Piagetian theory	III
105	Rowlands, S. (2010)	Mathematics education	Cultural-historical approach in teaching geometry	II

(continued)

(continued)

No.	Authors in the reference	Author's area of research	Context of the study	Level
106	Rowlands, S., Graham, T., & Berry, J. (2011)	Mathematics education	Paul Ernest's philosophy of mathematics educations	III
107	Rusanen, A.-M., & Pöyhönen, S. (2013)	Philosophy	Mechanisms of conceptual change	IV
108	Schmaus, W. (2008)	Philosophy of science	Social location in science	I
109	Schulz, R.M. (2009)	Science education	Philosophy of science education	III
110	Schumacher, A., & Reiners, C.S. (2013)	Chemistry education	Authentic learning	III
111	Shibley, I.V. (2003)	Philosophy of science	Newspapers and nature of science	II
112	Sievers, K.H. (1999)	Philosophy	Understanding observation	IV
113	Silverman, M.P. (1992)	Physics	Controversy in science	IV
114	Skordoulis, C.D. (2008)	Physics education	Worldviews & Marxism	II
115	Slezak, P. (1994)	Philosophy of science	Sociology of scientific knowledge	I
116	Smith, M.U., Siegel, H., & McInerney, J.D. (1995)	Biology education	Evolution & creationism controversy	III
117	Stafford, E. (2004)	Science education	Pendulum & scientific reasoning	II
118	Stolberg, T.L. (2009)	Science education	Religious education & evolution	II
119	Suchting, W.A. (1992)	Philosophy of science	Radical constructivism	III
120	Suchting, W.A. (1994)	Philosophy of science	Cultural significance of science	II
121	Suchting, W.A. (1995)	Philosophy of science	Nature of scientific thought	II
122	Szybek, P. (2002)	Science education	Scientific knowledge & human experience	II
123	Takacs, P., & Ruse, M. (2013)	Philosophy of biology	Philosophy of biology & its current status	III
124	Talanquer, V. (2013)	Chemistry education	Diversity in scientific thinking	III
125	Trumper, R. (2003)	Science education	Physics lab in a historical context	II
126	Uebel, T.E. (2004)	Philosophy of science	Education, enlightenment & positivism	III

(continued)

(continued)

No.	Authors in the reference	Author's area of research	Context of the study	Level
127	Vermeir, K. (2013)	History of science	Commodification of science	IV
128	Vesterinen, V.-M., Aksela, M., & Lavonen, J. (2013)	Chemistry education	Nature of science in school science textbooks	II
129	Wan, Z.H., Wong, S.L., & Zhan, Y. (2013)	Science education	Marxism and nature of science	III
130	Wong, S.L., Kwan, J., Hodson, D., Jung, B.H. W. (2009)	Science education	Nature of science & SARS	IV
131	Yasri, P., Arthur, S., Smith, M.U., & Mancy, R. (2013)	Science education	Science & religion	III

Notes:

1. In the case of more than one author, area of research refers to that of the first author. For a description of Levels of classification (I, II, III, IV and V) see Chap. 3.

Appendix 3

Articles from the *Journal of Research in Science Teaching* (Wiley Blackwell) evaluated in this study

Abd-El-Khalick, F., Waters, M., & Le, A.-P. (2008). Representations of nature of science in high school chemistry textbooks over the past four decades. *Journal of Research in Science Teaching, 45*(7), 835–855.

Akerson, V.L., Abd-El-Khalick, F., & Lederman, N.G. (2000). Influence of a reflective explicit activity-based approach on elementary teachers' conceptions of nature of science. *Journal of Research in Science Teaching, 37*(4), 295–317.

Akerson, V.L., Abd-El-Khalick, F., & McDuffie, A.R. (2006). One course is not enough: Preservice elementary teachers' retention of improved views of nature of science. *Journal of Research in Science Teaching, 43*(2), 194–213.

Akerson, V.L., & Volrich, M.L. (2006). Teaching nature of science explicitly in a first-grade internship setting. *Journal of Research in Science Teaching, 43*(4), 377–394.

Akerson, V.L., Buzzelli, C.A., & Donnelly, L.A. (2008). Early childhood teachers' views of nature of science: The influence of intellectual levels, cultural values, and explicit reflective teaching. *Journal of Research in Science Teaching, 45*(6), 748–770.

Akerson, V.L., Cullen, T.A., & Hanson, D.L. (2009). Fostering a community of practice through a professional development program to improve elementary teachers' views of nature of science and teaching practice. *Journal of Research in Science Teaching, 46*(10), 1090–1113.

Baker, D., & Leary, R. (1995). Letting girls speak out about science. *Journal of Research in Science Teaching, 32*(1), 3–27.

Barton, A.C. (1998). Teaching science with homeless children: Pedagogy, representation and identity. *Journal of Research in Science Teaching, 35*(4), 379–394.

Barton, A.C. (2001a). Capitalism, critical pedagogy, and urban science education: An interview with Peter McLaren. *Journal of Research in Science Teaching, 38*(8), 847–859.

Barton, A.C. (2001b). Science education in urban settings: Seeking new ways of praxis through critical ethnography. *Journal of Research in Science Teaching, 38*(8), 899–917.

Barton, A.C., & Yang, K. (2000). The culture of power and science education: Learning from Miguel. *Journal of Research in Science Teaching, 37*(8), 871–889.

Bartos, S.A., & Lederman, N.G. (2014). Teachers' knowledge structures for nature of science and scientific inquiry: Conceptions and classroom practice. *Journal of Research in Science Teaching, 51*(9), 1150–1184.

Bell, R.L., Blair, L.M., Crawford, B.A., & Lederman, N.G. (2003). Just do it? Impact of a science apprenticeship program on high school students' understandings of the nature of science and scientific inquiry. *Journal of Research in Science Teaching, 40*(5), 487–509.

Ben-Zvi Assaraf, O., & Orion, N. (2005). Development of system thinking skills in the context of earth system education. *Journal of Research in Science Teaching, 42*(5), 518–560.

Bev-Zvi Assaraf, O., & Orion, N. (2010). Four case studies, six years later: Developing system thinking skills in junior high school and sustaining them over time. *Journal of Research in Science Teaching, 47*(10), 1253–1280.

Bianchini, J.A., Cavazos, L.M., & Helms, J.V. (2000). From professional lives to inclusive practice: Science teachers and scientists' views of gender and ethnicity in science education. *Journal of Research in Science Teaching, 37*(6), 511–547.

Bianchini, J.A., & Colburn, A. (2000). Teaching the nature of science through inquiry to prospective elementary teachers: A tale of two researchers. *Journal of Research in Science Teaching, 37*(2), 177–209.

Bianchini, J.A., Hilton-Brown, B.A., & Breton, T.D. (2002). Professional development for university scientists around issues of equity and diversity: Investigating dissent within community. *Journal of Research in Science Teaching*, 39(8), 738–771.

Bianchini, J.A., & Solomon, E.M. (2003). Constructing views of science tied to issues of equity and diversity: A study of beginning science teachers. *Journal of Research in Science Teaching, 40*(1), 53–76.

Bismack, A.S., Arias, A.M., Davis, E.A., & Palincsar, A.S. (2015). Examining student work for evidence of teacher uptake of educative curriculum materials. *Journal of Research in Science Teaching, 52*(6), 816–846.

Boulton, A., & Panizzon, D. (1998). The knowledge explosion in science education: Balancing practical and theoretical knowledge. *Journal of Research in Science Teaching, 35*(5), 475–481.

Brickhouse, N. (2001). Embodying science: A feminist perspective on learning. *Journal of Research in Science Teaching, 38*(3), 282–295.

Briscoe, C. (1993). Using cognitive referents in making sense of teaching: A chemistry teacher's struggle to change assessment practices. *Journal of Research in Science Teaching, 30*(8), 971–987.

Brotman, J.S., & Moore, F.M. (2008). Girls and science: A review of four themes in the science education literature. *Journal of Research in Science Teaching, 45*(9), 971–1002.

Carter, L. (2008). Globalization and science education: The implications of science in the new economy. *Journal of Research in Science Teaching, 45*(5), 617–633.

Cavazos, L., Hazelwood, C.C., Howes, E.V., Kurth, L., Lane, P., Markham, L., Richmond, G., & Roth, K.J. (1998). Response to guest editorial: The WISE group: Connecting activism, teaching, and research. *Journal of Research in Science Teaching, 35*(4), 341–344.

Chen, S., Chang, W.-H., Lieu, S.-C., Kao, H.-L., Huang, M.-T., & Lin, S.-F. (2013). Development of an empirically based questionnaire to investigate young students' ideas about nature of science. *Journal of Research in Science Teaching, 50*(4), 408–430.

Chiappetta, E.L., Sethna, G.H., & Fillman, D.A. (1993). Do middle school life science textbooks provide a balance of scientific literacy themes? *Journal of Research in Science Teaching, 30*(7), 787–797.

Christodoulou, A., & Osborne, J. (2014). The science classroom as a site of epistemic talk: A case study of a teacher's attempts to teach science based on argument. *Journal of Research in Science Teaching, 51*(10), 1275–1300.

Cobern, W.W. (1994). Point: Belief, understanding, and the teaching of evolution. *Journal of Research in Science Teaching, 31*(5), 583–590.

Crawford, B., Zembal-Saul, C., Munford, D., & Friedrichsen, P. (2005). Confronting prospective teachers' ideas of evolution and scientific inquiry using technology and inquiry-based tasks. *Journal of Research in Science Teaching, 42*(6), 613–637.

Cronin, C., & Roger, A. (1999). Theorizing progress: Women in science, engineering, and technology in higher education. *Journal of Research in Science Teaching, 36*(6), 637–661.

Cross, R.T., & Price, R.F. (1996). Science teachers' social conscience and the role of controversial issues in the teaching of science. *Journal of Research in Science Teaching, 33*(3), 319–333.

DeBoer, G.E. (2000). Scientific literacy: Another look at its historical and contemporary mean-ings and its relationship to science education reform. *Journal of Research in Science Teaching, 37*(6), 582–601.

Dori, Y.J., & Herscovitz, O. (1999). Question-posing capability as an alternative evaluation method: Analysis of an environmental case study. *Journal of Research in Science Teaching, 36*(4), 411–430.

Driver, R. (1997). The application of science education theories: A reply to Stephen P. Norris and Tone Kvernbekk. *Journal of Research in Science Teaching, 34*(10), 1007–1018.

Duveen, J., & Solomon, J. (1994). The great evolution trial: Use of role-play in the classroom. *Journal of Research in Science Teaching, 31*(5), 575–582.

Ebenezer, J., Kaya, O.N., & Ebenezer, D.L. (2011). Engaging students in environmental research projects: Perceptions of fluency with innovative technologies and levels of scientific inquiry abilities. *Journal of Research in Science Teaching, 48*(1), 94–116.

Edmondson, K.M., & Novak, J.D. (1993). The interplay of scientific epistemological views, learning strategies, and attitudes of college students. *Journal of Research in Science Teaching, 30*(6), 547–559.

Eflin, J.T., Glennan, S., & Reisch, G. (1999). The nature of science: A perspective from the phi-losophy of science. *Journal of Research in Science Teaching, 36*(1), 107–116.

Feinstein, N.W. (2015). Education, communication, and science in the public sphere. *Journal of Research in Science Teaching, 52*(2), 145–163.

Florence, M.K., & Yore, L.D. (2004). Learning to write like a scientist: Coauthoring as an encul-turation task. *Journal of Research in Science Teaching, 41*(6), 637–668.

Fosnot, C.T. (1993). Rethinking science education: A defense of Piagetian constructivism. *Journal of Research in Science Teaching, 30*(9), 1189–1201.

Fusco, D., & Barton, A.C. (2001). Representing student achievements in science. *Journal of Research in Science Teaching, 38*(3), 337–354.

Fusco, D. (2001). Creating relevant science through urban planning and gardening. *Journal of Research in Science Teaching, 38*(8), 860–877.

Gazley, J.L., Remich, R., Naffziger-Hirsch, M.E., Keller, J., Campbell, P.B., & McGee, R. (2014). Beyond preparation: Identity, cultural capital, and readiness for graduate school in the biomedical sciences. *Journal of Research in Science Teaching, 51*(8), 1021–1048.

Good, R. (1993). Editorial: The slippery slopes of postmodernism. *Journal of Research in Science Teaching, 30*(5), 427.

Grindstaff, K., & Richmond, G. (2008). Learners' perceptions of the role of peers in a research experience: Implications for the apprenticeship process, scientific inquiry, and collaborative work. *Journal of Research in Science Teaching, 45*(2), 251–271.

Harding, P., & Hare, W. (2000). Portraying science accurately in classrooms: Emphasizing open-mindedness rather than relativism. *Journal of Research in Science Teaching, 37*(3), 225–236.

Hashweh, M.Z. (1996). Effects of science teachers' epistemological beliefs in teaching. *Journal of Research in Science Teaching, 33*(1), 47–63.

Havdala, R., & Ashkenazi, G. (2007). Coordination of theory and evidence: Effect of epistemolo-gical theories on students' laboratory practice. *Journal of Research in Science Teaching, 44*(8), 1134–1159.

Hildebrand, G.M. (1998). Disrupting hegemonic writing practices in school science: Contesting the right way to write. *Journal of Research in Science Teaching, 35*(4), 345–362.

Hogan, K., & Maglienti, M. (2001). Comparing the epistemological underpinnings of students' and scientists' reasoning about conclusions. *Journal of Research in Science Teaching, 38*(6), 663–687.

Howes, E.V. (1998). Connecting girls and science: A feminist teacher research study of a high school prenatal testing unit. *Journal of Research in Science Teaching, 35*(8), 877–896.

Hughes, G. (2000). Marginalization of socioscientific material in science-technology-society science curricula: Some implications for gender inclusivity and curriculum reform. *Journal of Research in Science Teaching, 37*(5), 426–440.

Jackson, D.F., Doster, E.C., Meadows, L., & Wood, T. (1995). Hearts and minds in the science classroom: The education of a confirmed evolutionist. *Journal of Research in Science Teaching, 32*(6), 585–611.

Jones, M.G., Tretter, T., Paechter, M., Kubasko, D., Bokinsky, A., Andre, T., & Negishi, A. (2007). Differences in African-American and European-American students' engagement with nanotechnology experiences: Perceptual position or assessment artifact? *Journal of Research in Science Teaching, 44*(6), 787–799.

Kawagley, A.O., Norris-Tull, D., & Norris-Tull, R.A. (1998). The indigenous worldview of Yupiaq culture: Its scientific nature and relevance to the practice and teaching of science. *Journal of Research in Science Teaching, 35*(2), 133–144.

Keig, P.F., & Rubba, P.A. (1993). Translation of representations of the structure of matter and its relationship to reasoning, gender, spatial reasoning, and specific prior knowledge. *Journal of Research in Science Teaching, 30*(8), 883–903.

Kelly, G.J., Chen, C., & Prothero, W. (2000). The epistemological framing of a discipline: Writing science in university oceanography. *Journal of Research in Science Teaching, 37*(7), 691–718.

Kittleson, J.M., & Southerland, S.A. (2004). The role of discourse in group knowledge construction: A case study of engineering students. *Journal of Research in Science Teaching, 41*(3), 267–293.

Kyle, W.C., Abell, S.A., Roth, W.-M., & Gallagher, J.J. (1992). Toward a mature discipline of science education. *Journal of Research in Science Teaching, 29*(9), 1015–1018.

Lather, P. (1998). Reaction to "Disrupting hegemonic writing practices in school science". *Journal of Research in Science Teaching, 35*(4), 363–364.

Liu, O.L., Lee, H.-S., & Linn, M.C. (2011). Measuring knowledge integration: Validation of four-year assessments. *Journal of Research in Science Teaching, 48*(9), 1079–1107.

Liu, O.L., Rios, J.A., Heilman, M., Gerard, L., & Linn, M.C. (2016). Validation of automated scoring of science assessments. *Journal of Research in Science Teaching, 53*(2), 215–233.

Lynch, S. (1994). Ability grouping and science education reform: Policy and research base. *Journal of Research in Science Teaching, 31*(2), 105–128.

Lynch, S. (1997). Novice teachers' encounter with national science education reform: Entanglements or intelligent interconnections? *Journal of Research in Science Teaching, 34*(1), 3–17.

Matthews, M.R. (1998). In defense of modest goals when teaching about the nature of science. *Journal of Research in Science Teaching, 35*(2), 161–174.

Mayberry, M. (1998). Reproductive and resistant pedagogies: The comparative roles of collaborative learning and feminist pedagogy in science education. *Journal of Research in Science Teaching, 35*(4), 443–459.

Nentwig, P., Roennebeck, S., Schoeps, K., Rumann, S., & Carstensen, C. (2009). Performance and levels of contextualization in a selection of OECD countries in PISA 2006. *Journal of Research in Science Teaching, 46*(8), 897–908.

Niaz, M. (2000). The oil drop experiment: A rational reconstruction of the Millikan-Ehrenhaft controversy and its implications for chemistry textbooks. *Journal of Research in Science Teaching, 37*(5), 480–508.

Nicolaidou, I., Kyza, E.A., Terzian, F., Hadjichambis, A., & Kafouris, D. (2011). A framework for scaffolding students' assessment of the credibility of evidence. *Journal of Research in Science Teaching, 48*(7), 711–744.

Norman, O. (1998). Marginalized discourses and literacy. *Journal of Research in Science Teaching, 35*(4), 365–374.

Norman, O., Ault, C.R., Bentz, B., & Meskimen, L. (2001). The black-white "achievement gap' as a perennial challenge of urban science education: A sociocultural and historical overview with implications for research and practice. *Journal of Research in Science Teaching, 38*(10), 1101–1114.

O'Loughlin, M. (1992). Rethinking science education: Beyond Piagetian constructivism toward a sociocultural model of teaching and learning. *Journal of Research in Science Teaching, 29*(8), 791–820.

Osborne, J., Collins, S., Ratcliffe, M., Millar, R., & Duschl, R. (2003). What "ideas-about-science" should be taught in school science? A Delphi study of the expert community. *Journal of Research in Science Teaching, 40*(7), 692–720.

Polman, J.L., & Gebre, E.H. (2015). Towards a critical appraisal of infographics as scientific inscriptions. *Journal of Research in Science Teaching, 52*(6), 868–893.

Richmond, G., Howes, E., Kurth, L., & Hazelwood, C. (1998). Connections and critique: Feminist pedagogy and science teacher education. *Journal of Research in Science Teaching, 35*(8), 897–918.

Ritchie, S.M., Tobin, K., & Hook, K.S. (1997). Teaching referents and the warrants used to test the viability of students' mental models: Is there a link? *Journal of Research in Science Teaching, 34*(3), 223–238.

Roth, W.-M., & Roychoudhury, A. (1993a). The development of science process skills in authentic contexts. *Journal of Research in Science Teaching, 30*(2), 127–152.

Roth, W.-M. (1993b). Heisenberg's uncertainty principle and interpretive research in science education. *Journal of Research in Science Teaching, 30*(7), 669–680.

Roth, W.-M. (1993). In the name of constructivism: Science education research and the construction of local knowledge. *Journal of Research in Science Teaching, 30*(7), 799–803.

Roth, W.-M., & Roychoudhury, A. (1994). Physics students' epistemologies and views about knowing and learning. *Journal of Research in Science Teaching, 31*(1), 5–30.

Roth, W.-M., & Lucas, K.B. (1997). From "truth" to "invented" reality: A discourse analysis of high school physics students' talk about scientific knowledge. *Journal of Research in Science Teaching, 34*(2), 145–179.

Roth, W.-M., & McGinn, M.K. (1998). >unDELETE science education:/lives/work/voices. *Journal of Research in Science Teaching, 35*(4), 399–421.

Sadler, T.D. (2004). Informal reasoning regarding socioscientific issues: A critical review of research. *Journal of Research in Science Teaching, 41*(5), 513–536.

Sadler, T.D., Amirshokoohi, A., Kazempour, M., & Allspaw, K.M. (2006). Socioscience and ethics in science classrooms: Teacher perspectives and strategies. *Journal of Research in Science Teaching, 43*(4), 353–376.

Schroeder, C.M., Scott, T.P., Tolson, H., Huang, T.-Y., & Lee, Y.-H. (2007). A meta-analysis of national research: Effects of teaching strategies on student achievement in science in the United States. *Journal of Research in Science Teaching, 44*(10), 1436–1460.

Sencar, S., & Eryilmaz, A. (2004). Factors mediating the effect of gender on ninth-grade Turkish students' misconceptions concerning electric circuits. *Journal of Research in Science Teaching, 41*(6), 603–616.

Shanahan, M.-C., & Nieswandt, M. (2011). Science student role: Evidence of social structural norms specific to school science. *Journal of Research in Science Teaching, 48*(4), 367–395.

Showers, D.E., & Shrigley, R.L. (1995). Effects of knowledge and persuasion on high-school students' attitudes toward nuclear power plants. *Journal of Research in Science Teaching, 32*(1), 29–43.

Shumba, O., & Glass, L.W. (1994). Perceptions of coordinators of college freshman chemistry regarding selected goals and outcomes of high school chemistry. *Journal of Research in Science Teaching, 31*(4), 381-392.

Siegel, M.A., & Ranney, M.A. (2003). Developing the changes in attitude about the relevance of science (CARS) questionnaire and assessing two high school science classes. *Journal of Research in Science Teaching, 40*(8), 757–775.

Smith, C.L., & Wenk, L. (2006). Relations among three aspects of first-year college students' epistemologies of science. *Journal of Research in Science Teaching, 43*(8), 747–785.

Smith, M.U. (1994). Counterpoint: Belief, understanding, and the teaching of evolution. *Journal of Research in Science Teaching, 31*(5), 591–597.

Snyder, V.L., & Broadway, F.S. (2004). Queering high school biology textbooks. *Journal of Research in Science Teaching, 41*(6), 617–636.

Staver, J.R. (1995). Scientific research and oncoming vehicles: Can radical constructivists embrace one and dodge the other? *Journal of Research in Science Teaching, 32*(10), 1125–1128.

Tomas, L., Ritchie, S.M., & Tones, M. (2011). Attitudinal impact of hybridized writing about a socioscientific issue. *Journal of Research in Science Teaching, 48*(8), 878–900.

Tsui, C.-Y., & Treagust, D.F. (2007). Understanding genetics: Analysis of secondary students' conceptual status. *Journal of Research in Science Teaching, 44*(2), 205–235.

Van Eijck, M., & Roth, W.-M. (2011). Cultural diversity in science education through *novelization*: Against the *epicization* of science and cultural centralization. *Journal of Research in Science Teaching, 48*(7), 824–847.

Venville, G. (2004). Young children learning about living things: A case study of conceptual change from ontological and social perspectives. *Journal of Research in Science Teaching, 41*(5), 449–480.

Verma, G., Puvirajah, A., & Webb, H. (2015). Enacting acts of authentication in a robotics competition: An interpretivist study. *Journal of Research in Science Teaching, 52*(3), 268–295.

Warren, B., Ballenger, C., Ogonowski, M., Roseberry, A.S., & Hudicourt-Barnes, J. (2001). Rethinking diversity in learning science: The logic of everyday sense-making. *Journal of Research in Science Teaching, 38*(5), 529–552.

Wenner, J.A., & Settlage, J. (2015). School leader enactments of the structure/agency dialectic via buffering. *Journal of Research in Science Teaching, 52*(4), 503–515.

Wilson, R.E., & Kittleson, J. (2013). Science as a classed and gendered endeavor: Persistence of two white female first-generation college students within an undergraduate science context. *Journal of Research in Science Teaching, 50*(7), 802–825.

Yerrick, R.K. (2000). Lower track science students' argumentation and open inquiry instruction. *Journal of Research in Science Teaching, 37*(8), 807–838.

Yore, L.D., Hand, B.M., & Florence, M.K. (2004). Scientists' views of science, models of writing, and science writing practices. *Journal of Research in Science Teaching, 41*(4), 338–369.

Zembylas, M. (2002). Constructing genealogies of teachers' emotions in science teaching. *Journal of Research in Science Teaching, 39*(1), 79–103.

Zoller, U. (1999). Scaling-up of higher-order cognitive skills-oriented college chemistry teaching: An action-oriented research. *Journal of Research in Science Teaching, 36*(5), 583–596.

Appendix 4

Distribution of articles (*Journal of Research in Science Teaching*) according to author's area of research, context of the study and level (classification)

No.	Authors in the reference	Author's area of research	Context of the study	Level
1	Abd-El-Khalick, F., Waters, M., & Le, A.-P. (2008)	Science education	Nature of science in chemistry textbooks	IV
2	Akerson, V.L., Abd-El-Khalick, F., & Lederman, N.G. (2000)	Science education	Teachers' conceptions of nature of science	III
3	Akerson, V.L., Morrison, J.A., & McDuffie, A.R. (2006)	Science education	Teachers' conceptions of nature of science	III
4	Akerson, V.L., & Volrich, M.L. (2006)	Science education	Teaching nature of science	III
5	Akerson, V.L., Buzzelli, C.A., & Donnelly, L.A. (2008)	Science education	Teachers' views of nature of science	III
6	Akerson, V.L., Cullen, T.A., & Hanson, D.L. (2009)	Science education	Teachers' views of nature of science	III
7	Baker, D., & Leary, R. (1995)	Science education	Science as a career for women	III
8	Barton, A.C. (1998)	Science education	Pedagogy, representation and identity	III
9	Barton, A.C., & Yang, K. (2000)	Science education	Culture of power and science education	III

(continued)

(continued)

No.	Authors in the reference	Author's area of research	Context of the study	Level
10	Barton, A.C. (2001a)	Science education	Capitalism, critical pedagogy and science education	III
11	Barton, A.C. (2001b)	Science education	Critical ethnography and science education	III
12	Bartos, S.A., & Lederman, N.G. (2014)	Science education	Teachers' knowledge structures for nature of science	II
13	Bell, R.L., Blair, L.M., Crawrford, B.A., & Lederman, N.G. (2003)	Science education	Science apprenticeship and nature of science	III
14	Ben-Zvi Assaraf, O., & Orion, N. (2005)	Science education	Earth system education	II
15	Ben-Zvi Assaraf, O., & Orion, N. (2010)	Science education	Developing system thinking skills	II
16	Bianchini, J.A., & Colburn, A. (2000)	Science education	Teaching nature of science to elementary teachers	III
17	Bianchini, J.A., Cavazos, L.M., Helms, J.V. (2000)	Science education	Gender and ethnicity in science education	III
18	Bianchini, J.A., Hilton-Brown, B.A., & Breton, T.D. (2002)	Science education	Dissent within community	III
19	Bianchini, J.A., Solomon, E.M. (2003)	Science education	Nature of science, equity and diversity	III
20	Bismack, A.S., Arias, A.M., Davis, E.A., & Palincsar, A.S. (2015)	Science education	Teacher uptake of educative curriculum materials	II
21	Boulton, A., & Panizzon, D. (1998)	Ecosystem management	Balancing practical and theoretical knowledge	III
22	Brickhouse, N. (2001)	Science education	Feminist perspective on learning	III
23	Briscoe, C. (1993)	Science education	Assessment practices	III
24	Brotman, J.S., & Moore, F.M. (2008)	Science education	Gender and science education	III
25	Carter, L. (2008)	Science education	Globalization, science and science education	III
26	Cavazos, L., et al. (1998)	Science education	Feminism and science education	III
27	Chen, S., et al. (2013)	Science education	Students' ideas about nature of science	IV
28	Chiappetta, E.L., Sethna, G.H., & Fillman, D.A. (1993)	Science education	Scientific literacy themes in textbooks	II

(continued)

(continued)

No.	Authors in the reference	Author's area of research	Context of the study	Level
29	Christodoulou, A., & Osborne, J. (2014)	Science education	Teaching science based on arguments	II
30	Cobern, W.W. (1994)	Science education	Belief, understanding and teaching of evolution	III
31	Crawford, B.A., Zembal-Saul, C., Munford, D., & Friedrichsen, P. (2005)	Science education	Confronting teachers' ideas of evolution	III
32	Cronin, C., & Roger, A. (1999)	Education	Women in science, engineering and technology	III
33	Cross, R.T., & Price, R.F. (1996)	Science education	Role of controversial issues	III
34	DeBoer, G.E. (2000)	Science education	Scientific literacy and science education reform	II
35	Dori, Y.J., & Herscovitz, O. (1999)	Science education	Question-posing capability	II
36	Driver, R. (1997)	Science education	Science education theories	III
37	Duveen, J., & Solomon, J. (1994)	Science education	Teaching evolution in the classroom	III
38	Ebenezer, J., Kaya, O.N., & Ebenezer, D.L. (2011)	Science education	Engaging students in environmental research	III
39	Edmondson, K.M., & Novak, J.D. (1993)	Science education	Students' epistemological views and learning strategies	IV
40	Eflin, J.T., Glennan, S., & Reisch, G. (1999)	Philosophy of science	Nature of science	II
41	Feinstein, N.W. (2015)	Science education	Science in the public sphere	I
42	Florence, M.K., & Yore, L.D. (2004)	Writing and editing	Learning to write like a scientist	II
43	Fosnot, C.T. (1993)	Teacher education	Piagetian constructivism	III
44	Fusco, D., & Barton, A.C. (2001)	Science education	Student achievement in science	III
45	Fusco, D. (2001)	Science education	Creating relevant science	III
46	Gazley et al. (2014)	Medicine	Graduate school and biomedical sciences	II
47	Germann, P.J., Aram, R., & Burke, G. (1996)	Science education	Science process skills	II

(continued)

(continued)

No.	Authors in the reference	Author's area of research	Context of the study	Level
48	Good, R. (1993)	Science education	Postmodernism and science education	III
49	Grindstaff, K., & Richmond, G. (2008)	Science education	Role of peers in research	I
50	Harding, P., & Hare, W. (2000)	Science education	Open-mindedness versus relativism	IV
51	Hashweh, M.Z. (1996)	Science education	Epistemological beliefs of science teachers	III
52	Havdala, R., & Ashkenazi, G. (2007)	Science education	Coordination of theory and evidence	III
53	Hildebrand, G.M. (1998)	Science education	Hegemonic writing practices in school science	III
54	Hogan, K., & Maglienti, M. (2001)	Science education	Underpinnings of students' and scientists' reasoning	III
55	Howes, E.V. (1998)	Teacher education	Feminism and prenatal testing	III
56	Hughes, G. (2000)	Science education	Marginalization of socio-scientific issues	IV
57	Jackson, D.F., Doster, E.C., Meadows, L., & Wood, T. (1995)	Science education	Education of a confirmed evolutionist	III
58	Jones, M.G., et al. (2007)	Science education	Students' engagement with nanotechnology	II
59	Kawagley, A.O., Norris-Tull, D., & Norris-Tull, R.A. (1998)	Science education	Indigenous worldview of Yupiaq culture	III
60	Keig, P.F., & Rubba, P.A. (1993)	Science education	Translation of representations of structure of matter	II
61	Kelly, G.J., Chen, C., & Prothero, W. (2000)	Science education	Epistemological framing of oceanography	II
62	Kittleson, J.M., & Southerland, S.A. (2004)	Science education	Role of discourse and knowledge construction	III
63	Kyle, W.C., Abell, S.K., Roth, W.-M., & Gallagher, J.J. (1992).	Science education	Science education as a mature discipline	III
64	Lather, P. (1998)	Feminist ethnography	Hegemonic writing practices in school science	III
65	Liu, O.L., Lee, H.-S., & Linn, M.C. (2011)	Educational assessment	Measuring knowledge integration	II

(continued)

(continued)

No.	Authors in the reference	Author's area of research	Context of the study	Level
66	Liu, O.L., et al. (2016)	Educational assessment	Validation of automated scoring	II
67	Lynch, S. (1994)	Science education	Ability grouping and science education reform	II
68	Lynch, S. (1997)	Science education	Teachers and national science education reform	II
69	Matthews, M.R. (1998)	Science education	Teaching about nature of science	I
70	Mayberry, M. (1998)	Women's studies	Feminist pedagogy in science education	III
71	Nentwig, P., et al. (2009)	Science education	Performance of OECD countries in PISA	II
72	Niaz, M. (2000)	Science education	Presentation of oil drop experiment in chemistry textbooks	III
73	Nicolaidou et al. (2011)	Communication & internet studies	Scaffolding students' assessment	II
74	Norman, O. (1998)	Science education	Marginalized discourses	III
75	Norman, O. et al. (2001)	Science education	The black-white achievement gap in urban science education	III
76	O'Loughlin, M. (1992)	Teacher education	Sociocultural model of teaching and learning	III
77	Osborne, J., et al. (2003)	Science education	Ideas-about-science and school science	II
78	Polman, J.L., & Gebre, E.H. (2015)	Science education	Infographics as scientific inscriptions	II
79	Richmond, G., et al. (1998)	Teacher education	Feminist pedagogy and science teacher education	III
80	Ritchie, S.M., Tobin, K., & Hook, K.S. (1997)	Science education	Viability of students' mental models	III
81	Roth, W.-M., & Roychoudhury, A. (1993)	Science education	Science process skills in authentic contexts	II
82	Roth, W.-M. (1993)a	Science education	Heisenberg's uncertainty principle and science education	III
83	Roth, W.-M. (1993)b	Science education	Constructivism and science education research	III

(continued)

(continued)

No.	Authors in the reference	Author's area of research	Context of the study	Level
84	Roth, W.-M., & Roychoudhury, A. (1994)	Science education	Physics students' epistemologies	II
85	Roth, W.-M., & Lucas, K.B. (1997)	Science education	Physics students' talk about scientific knowledge	III
86	Roth, W.-M., & McGinn, M.K. (1998)	Science education	Grading practices and science education	III
87	Sadler, T.D. (2004)	Science education	Informal reasoning and socioscientific issues	II
88	Sadler, T.D. et al. (2006)	Science education	Teacher perspectives on socioscience and ethics	III
89	Schroeder, C.M. (2007)	Science education	Teaching strategy and student achievement	II
90	Sencar, S., & Eryilmaz, A. (2004)	Science education	Gender and misconceptions concerning electric circuits	II
91	Shanahan, M.-C., & Nieswandt, M. (2011)	Science education	Social structural norms of school science	III
92	Showers, D.E., & Shrigley, R.L. (1995)	Science education	Students' attitudes toward nuclear power plants	II
93	Shumba, O., & Glass, L. W. (1994)	Science education	Perceptions of high school chemistry coordinators	II
94	Siegel, M.A., & Ranney, M.A. (2003)	Science education	Changes in attitude about the relevance of science	II
95	Smith, C.L., & Wenk, L. (2006)	Psychology	College students' epistemologies of science	II
96	Smith, M.U. (1994)	Science education	Belief, understanding, and the teaching of evolution	III
97	Snyder, V.L., & Broadway, F.S. (2004)	Science education	Queer theory and biology textbooks	III
98	Staver, J.R. (1995)	Science education	Understanding radical constructivism	III
99	Tomas, L., Ritchie, S. M., & Tones, M (2011)	Science education	Hybridized writing about socioscientific issues	III
100	Tsui, C.-Y., & Treagust, D.F. (2007)	Science education	Rigor of qualitative research	III

(continued)

(continued)

No.	Authors in the reference	Author's area of research	Context of the study	Level
101	van Eijck, M., & Roth, W.-M. (2011)	Science education	Cultural diversity in science education	I
102	Venville, G. (2004)	Science education	Young children learning about living things	III
103	Verma, G., Puvirajah, A., & Webb, H. (2015)	Science education	Authentication in a robotics competition	III
104	Warren, B., et al. (2001)	Science education	Rethinking diversity in learning science	III
105	Wenner, J.A., & Settlage, J. (2015)	Science education	School leadership and structure/agency dialectic	III
106	Wilson, R.E., & Kittleson, J. (2013)	Science education	Science as a class and gendered endeavor	III
107	Yerrick, R.K. (2000)	Science education	Students' argumentation and inquiry instruction	III
108	Yore, L.D., Hand, B,M., & Florence, M.K. (2004)	Science education	Scientists' writing practices	III
109	Zembylas, M. (2002)	Science education	Teachers' emotions in science teaching	III
110	Zoller, U. (1999)	Science education	Higher-order cognitive skills in teaching chemistry	III

Notes:
1. In the case of more than one author, area of research refers to that of the first author.
2. For a description of Levels of classification (I, II, III, IV and V) see Chap. 3

Appendix 5

Articles from the *International Handbook of Research in History, Philosophy and Science Teaching* (Springer) evaluated in this study

Galili, I. (2014). Teaching optics: A historico-philosophical perspective. In M.R. Matthews (Ed.), *International Handbook of Research in History, Philosophy and Science Teaching* (Vol. I, pp. 97–128). Dordrecht: Springer.

Glas, E. (2014). A role for quasi-empiricism in mathematics education. In M.R. Matthews (Ed.), *International Handbook of Research in History, Philosophy and Science Teaching* (Vol. I, pp. 731–753). Dordrecht: Springer.

Horsthemke, K., & Yore, L.D. (2014). Challenges of multiculturalism in science education: Indigenisation, internationalism, and *transkulturalität*. In M.R. Matthews (Ed.), *International Handbook of Research in History, Philosophy and Science Teaching* (Vol. III, pp. 1759–1792). Dordrecht: Springer.

Mackenzie, J., Good, R., & Brown, J.R. (2014). Postmodernism and science education: An appraisal. In M.R. Matthews (Ed.), *International Handbook of Research in History, Philosophy and Science Teaching* (Vol. II, pp. 1057–1086). Dordrecht: Springer.

McCarthy, C.L. (2014). Cultural studies in science education: Philosophical considerations. In M.R. Matthews (Ed.), *International Handbook of Research in History, Philosophy and Science Teaching* (Vol. III, pp. 1927–1964).

Reiss, M.J. (2014). What significance does Christianity have for science education? In M.R. Matthews (Ed.), *International Handbook of Research in History, Philosophy and Science Teaching* (Vol. II, pp. 1637–1662). Dordrecht: Springer.

Schulz, R.M. (2014). Philosophy of education and science education: A vital but underdeveloped relationship. In M.R. Matthews (Ed.), *International Handbook of Research in History, Philosophy and Science Teaching* (Vol. II, pp. 1259–1316). Dordrecht: Springer.

Taber, K.S. (2014). Methodological issues in science education research: A perspective from the philosophy of science. In M.R. Matthews (Ed.), *International Handbook of Research in History, Philosophy and Science Teaching* (Vol. III, pp. 1839–1893). Dordrecht: Springer.

Appendix 6

Distribution of articles (*International Handbook of Research in History, Philosophy and Science Teaching*) according to author's area of research, context of the study and level (classification)

No.	Authors in the reference	Author's area of research	Context of the study	Level
1	Galili, I. (2014)	Science education	Teaching optics	III
2	Glas, E. (2014)	Mathematics	Mathematics education	III
3	Horsthemke, K., & Yore, L.D. (2014)	Education	Multiculturalism	II
4	Mackenzie, J., Good, R., & Brown, J.R. (2014)	Education	Feminism and science	III
5	McCarthy, C.L. (2014)	Philosophy of education	Cultural studies	IV
6	Reiss, M.J. (2014)	Science education	Nature of science	II
7	Schulz, R.M. (2014)	Science education	Philosophy of science	II
8	Taber, K.S. (2014)	Science education	Research methodology	II

Notes:
1. In the case of more than one author, area of research refers to that of the first author.
2. For a description of Levels (I, II, III, IV & V) see Chap. 3.

Appendix 7

Articles from *Encyclopedia of Science Education* (Springer) evaluated in this study

Alsop, S. (2015). Affect in learning science. In R. Gunstone (Ed.), *Encyclopedia of Science Education* (pp. 19–24). Heidelberg: Springer.

Brickhouse, N. (2015). Gender. In R. Gunstone (Ed.), *Encyclopedia of Science Education* (pp. 440–441). Heidelberg: Springer.

Cavas, B. (2015). Values. In R. Gunstone (Ed.), *Encyclopedia of Science Education* (pp. 1089–1090). Heidelberg: Springer.

Corrigan, D. (2015). Curriculum and values. In R. Gunstone (Ed.), *Encyclopedia of Science Education* (pp. 256–258).

Fischler, H. (2015). Bildung. In R. Gunstone (Ed.), *Encyclopedia of Science Education* (pp. 118–122).

Irzik, G. (2015). Values and Western science knowledge. In R. Gunstone (Ed.), *Encyclopedia of Science Education* (pp. 1093–1096). Heidelberg: Springer.

Reiss, M.J. (2015). Religious education and science education. In R. Gunstone (Ed.), *Encyclopedia of Science Education* (pp. 831–834). Heidelberg: Springer.

Robinson, D. (2015). Broadcast media. In R. Gunstone (Ed.), *Encyclopedia of Science Education* (pp. 135–138). Heidelberg: Springer.

Rudolph, J.L. (2015). Science studies. In R. Gunstone (Ed.), *Encyclopedia of Science Education* (pp. 914–917). Heidelberg: Springer.

Scantlebury, K. (2015). Sociocultural perspectives and gender. In R. Gunstone (Ed.), *Encyclopedia of Science Education* (pp. 983–985). Heidelberg: Springer.

Stewart, G.M. (2015). Ethnoscience. In R. Gunstone (Ed.), *Encyclopedia of Science Education* (pp. 401–402). Heidelberg: Springer.

Taylor, P.C. (2015). Constructivism. In R. Gunstone (Ed.), *Encyclopedia of Science Education* (pp. 218–224). Heidelberg: Springer.

Appendix 8

Distribution of articles (*Encyclopedia of Science Education*) according to author's area of research, context of the study and level (classification)

No.	Authors in the reference	Author's area of research	Context of the study	Level
1	Alsop, S. (2015)	Science education	Affect in learning science	IV
2	Brickhouse, N. (2015)	Science education	Gender	III
3	Cavas, B. (2015)	Science education	Values	II
4	Corrigan, D. (2015)	Education	Curriculum and values	III
5	Fischler, H. (2015)	Science education	Bildung	II
6	Irzik, G. (2015)	Philosophy of science	Values and Western science	III
7	Reiss, M.J. (2015)	Science education	Religious education	II
8	Robinson, D. (2015)	Science education	Broadcast media	II
9	Rudolph, J.L. (2015)	Science education	Science studies	II
10	Scantlebury, K. (2015)	Science education	Sociocultural perspectives & gender	II
11	Stewart, G.M. (2015)	Science education	Ethnoscience	III
12	Taylor, P.C. (2015)	Science education	Constructivism	IV

Note:
1. For a description of Levels (I, II, III, IV & V) see Chap. 3.

Appendix 9

General chemistry textbooks (published in USA) evaluated in this study ($n = 60$)

Armstrong, J. (2012). *General, organic and biochemistry: An applied approach.* Belmont, CA: Brooks/Cole.

Atkins, P., & Jones, L. (2002). *Chemical principles: The quest for insight* (2nd ed.). New York: Freeman.

Atkins, P., & Jones, L. (2008). *Chemical principles: The quest for insight* (4th ed.). New York: Freeman.

Bettelheim, F.A., Brown, W.H., Campbell, M.K., & Farrell, S.O. (2010). *Introduction to general, organic and biochemistry* (9th ed.). Belmont, CA: Brooks/Cole.

Bishop, M. (2002). *An introduction to chemistry.* San Francisco: Benjamin Cummings.

Blei, I., & Odian, G. (2006). *General, organic and biochemistry: Connecting chemistry to your life* (2nd ed.). New York: W.H. Freeman.

Brady, J.E., & Humiston, G. (1996). *General chemistry: Principles and structure* (Spanish ed.). New York: Wiley.

Brady, J.E., Russell, J., & Holum, J. (2000). *Chemistry: The study and its changes* (3rd ed.). New York: Wiley.

Brady, J.E., & Senese, F.A. (2009). *Chemistry: Matter and its changes* (5th ed.). Hoboken, NJ: Wiley.

Brown, L.S., & Holme, T.A. (2011). *Chemistry for engineering students* (2nd ed.). Belmont, CA: Brooks/Cole.

Brown, T.L., LeMay, H.E., & Bursten, B. (1997). *Chemistry: The central science* (7th ed., Spanish). Englewood Cliffs, NJ: Prentice Hall.

Brown, T.L., LeMay, H.E., Bursten, B.E., & Murphy, C.J. (2009). *Chemistry: The central science* (11th ed., Spanish). Englewood Cliffs, NJ: Prentice Hall.

Brown, T.L., LeMay, H.E., Bursten, B.E., Murphy, C.J., & Woodward, P. (2014). *Chemistry: The central science* (12th ed.). Essex, UK: Pearson International Education edition.

Burns, R. (1995). *Fundamentals of chemistry* (2nd ed., Spanish). Englewood Cliffs, NJ: Prentice Hall.

Chang, R. (1998). *Chemistry* (6th ed., Spanish). New York: McGraw Hill.

Chang, R. (2010). *Chemistry* (10th ed., Spanish). New York: McGraw Hill.

Cracolice, M.S., & Peters, E. (2016). *Introductory chemistry: An active learning approach* (6th ed.). Boston, MA: Cengage Learning.

Daub, G.W., & Seese, W. (1996). *Basic chemistry* (8th ed., Spanish). Englewood Cliffs: Prentice Hall.

Denniston, K.J., Topping, J.J., & Caret, R.L. (2011). *General, organic and biochemistry* (7th ed.). New York: McGraw Hill.

Dickson, T. (2000). *Introduction to chemistry* (8th ed.). New York: Wiley.

Ebbing, D.D. (1996). *General chemistry* (5th ed., Spanish). New York: McGraw Hill.

Ebbing, D.D., & Gammon, S.D. (2013). *General chemistry* (10th ed.). Belmont, CA: Brooks/Cole.

Ebbing, D.D., & Gammon, S.D. (2017). *General chemistry* (11th ed.). Boston, MA: Cengage Learning.

Ellis, A.B., Geselbracht, M.J., Johnson, B.J., Lisensky, G.C., & Robinson, W.R. (1993). *Teaching general chemistry: A materials science companion.* Washington, D.C.: American Chemical Society.

Frost, L., Deal, T., & Timberlake, K.C. (2011). *General, organic and biological chemistry: An integrated approach.* Upper Saddle River, NJ: Prentice Hall.

Garoutte, M.P., & Mahoney, A.B. (2015). *Introductory chemistry: A guided inquiry.* Hoboken, NJ: Wiley.

Goldberg, D.E. (2001). *Fundamentals of chemistry* (3rd ed.). New York: McGraw Hill.

Hein, M. (1990). *Foundations of college chemistry.* Belmont, CA: Brooks/Cole.

Hein, M., & Arena, S. (1997). *Foundations of college chemistry.* Belmont, CA: Brooks/Cole.

Hill, J., & Petrucci, R. (1999). *General chemistry: An integrated approach* (2nd ed.). Upper Saddle River, NJ: Prentice Hall.

Joesten, M.D., Castellion, M.E., & Hogg, J.L. (2007). *The world of chemistry: Essentials* (4th ed.). Belmont, CA: Brooks/Cole.

Joesten, M. D., Johnstone, D.O., Netterville, J.T., & Wood, J.L. (1991). *World of chemistry.* Philadelphia: Saunders.

Jones, L., & Atkins, P. (2000). *Chemistry: Molecules, matter and change* (4th ed.). New York: Freeman.

Kotz, J.C., Treichel, P.M., Townsend, J.R., & Treichel, D.A. (2015). *Chemistry and chemical reactivity* (9th ed.). Stamford, CT: Cengage Learning.

Malone, L.J. (2001). *Basic concepts of chemistry* (6th ed.). New York: Wiley.

Malone, L.J., & Dolter, T.O. (2013). *Basic concepts of chemistry* (9th ed.). Hoboken, NJ: Wiley.

Masterton, W.L., Hurley, C.N., & Neth, E.J. (2012). *Chemistry: Principles and reactions* (7th ed.). Belmont, CA: Brooks/Cole.

McMurry, J., Castellion, M.E., & Ballantine, D.S. (2007). *Fundamentals of general, organic and biological chemistry* (5th ed.). Upper Saddle River, NJ: Prentice Hall.

McMurry, J., & Fay, R. (2001). *Chemistry* (3rd ed.). Upper Saddle River, NJ: Prentice Hall.

Mcquarrie, D.A., Rock, P.A., & Gallogly, E.B. (2011). *General chemistry* (4th ed.). Mill Valley, CA: University Science Books.

Moore, J.W., Stanitski, C.L., & Jurs, P.C. (2002). *Chemistry: The molecular science.* Orlando, FL: Harcourt College Publishers.

Moore, J.W., Stanitski, C.L., & Jurs, P.C. (2011). *Chemistry: The molecular science* (4th ed.). Belmont, CA: Brooks/Cole.

Olmsted, J.A., & Williams, G.M. (2006). *Chemistry* (4th ed.). Hoboken, NJ: Wiley.

Oxtoby, D.W., Gillis, H.P., & Campion, A. (2012). *Principles of modern chemistry* (7th ed.). Belmont, CA: Brooks/Cole.

Oxtoby, D.W., Nachtrieb, N., & Freeman, W. (1990). *Chemistry: Science of change* (2nd ed.). Philadelphia: Saunders.

Raymond, K.W. (2010). *General, organic and biological chemistry: An integrated approach* (3rd ed.). Hoboken, NJ: Wiley.

Russo, S., & Silver, M. (2002). *Introductory chemistry* (2nd ed.). San Francisco: Benjamin Cummings.

Seager, S.L., Slabaugh, M.R. (2011). *Chemistry for today: General, organic and biochemistry.* Belmont, CA: Brooks/Cole.

Silberberg, M. (2000). *Chemistry: The molecular nature of matter and change* (2nd ed.). New York: McGraw Hill.

Spencer, J.N., Bodner, G.M., & Rickard, L.H. (1999). *Chemistry: Structure and dynamics*. New York: Wiley.

Spencer, J.N., Bodner, G.M., & Rickard, L.H. (2012). *Chemistry: Structure and dynamics* (5th ed.). Hoboken, NJ: Wiley.

Stoker, H.S. (2010). *General, organic and biological chemistry* (5th ed.). Belmont, CA: Brooks/Cole.

Stoker, H.S. (2016). *General, organic and biological chemistry* (7th ed.). Boston, MA: Cengage Learning.

Timberlake, K.C. (2010). *General, organic and biological chemistry: Structures of life* (3rd ed.). Upper Saddle River, NJ: Prentice Hall.

Tro, N.J. (2008). *Chemistry: A molecular approach*. Upper Saddle River, NJ: Prentice Hall.

Umland, J., & Bellama, J. (1999). *General chemistry* (3rd ed.). Pacific Grove, CA: Brooks/Cole.

Whitten, K.W., Davis, R.E., Peck, M.L., & Stanley, G.G. (2010). *Chemistry* (10th ed.). Belmont, CA: Brooks/Cole.

Zumdahl, S.S., & Decoste, D.J. (2015). *Introductoy chemistry: A foundation* (8th ed.). Stamford, CT: Cengage Learning.

Zumdahl, S.S., & Zumdahl, S.A. (2007). *Chemistry* (7th ed.). Boston, MA: Houghton Mifflin Co.

Zumdahl, S.S., & Zumdahl, S.A. (2014). *Chemistry* (9th ed.). Belmont, CA: Brooks/Cole.

Appendix 10

Evaluation of general chemistry textbooks published in USA ($n = 60$)

No.	Textbook Points[b]	Criteria[a] 1	2	3	4	5	
1	Armstrong (2012)	N	N	N	N	N	0
2	Atkins & Jones (2002)	N	M	S	S	M	6
3	Atkins & Jones (2008)	N	M	S	S	S	7
4	Bettelheim et al (2010)	N	S	N	N	M	3
5	Bishop (2002)	N	N	N	N	N	0
6	Blei & Odian (2006)	N	M	N	N	N	1
7	Brady & Humiston (1996)	N	N	N	N	N	0
8	Brady, Russell & Holum (2000)	N	M	N	N	N	1
9	Brady & Senese (2009)	N	S	S	N	S	6
10	Brown & Holme (2011)	N	S	N	N	S	4
11	Brown, LeMay & Bursten (1997)	N	M	N	N	N	1
12	Brown, LeMay, Bursten & Murphy (2009)	N	M	M	N	S	4
13	Brown et al. (2014)	N	M	N	N	S	3
14	Burns (1995)	N	N	N	N	N	0
15	Chang (1998)	N	M	N	N	S	3
16	Chang (2010)	N	M	N	N	S	3
17	Cracolice & Peters (2016)	N	S	M	M	S	6
18	Daub & Seese (1996)	N	N	N	N	N	0
19	Denniston, Topping & Caret (2011)	N	S	M	N	N	3

(continued)

(continued)

20	Dickson (2000)	N	N	S	N	N	2
21	Ebbing (1996)	N	M	M	N	N	2
22	Ebbing & Gammon (2013)	N	M	M	M	N	3
23	Ebbing & Gammon (2017)	N	M	M	S	M	5
24	Ellis et al (1993)	N	N	M	M	M	3
25	Frost, Deal & Timberlake (2011)	N	N	N	N	N	0
26	Garouttte & Mahoney (2015)	N	N	N	N	N	0
27	Goldberg (2001)	N	N	N	N	N	0
28	Hein (1990)	N	M	N	N	N	1
29	Hein & Arena (1997)	N	M	M	M	N	3
30	Hill & Petrucci (1999)	N	M	S	N	M	4
31	Joesten, Castellion & Hogg (2007)	N	N	S	M	N	3
32	Joesten et al (1991)	M	S	M	N	N	4
33	Jones & Atkins (2000)	N	N	N	N	S	2
34	Kotz, et al (2015)	S	S	S	N	M	7
35	Malone (2001)	N	N	N	N	N	0
36	Malone & Dolter (2013)	N	M	N	N	N	1
37	Masterton, Hurley & Neth (2012)	N	N	N	N	N	0
38	McMurry, Castellion & Ballantine (2007)	N	N	S	N	N	2
39	McMurry & Fay (2001)	N	N	S	N	M	3
40	Mcquarrie, Rock & Gallogly (2011)	N	N	N	N	N	0
41	Moore, Stanitski & Jurs (2002)	N	N	S	N	N	2
42	Moore, Stanitski & Jurs (2011)	N	N	S	S	S	6
43	Olmsted & Williams (2006)	N	M	S	S	S	7
44	Oxtoby, Gillis & Campion (2012)	N	N	S	S	M	5
45	Oxtoby, Nachtrieb & Freeman (1990)	N	N	S	N	N	2
46	Raymond (2010)	N	M	N	N	N	1
47	Russo & Silver (2002)	N	N	M	N	N	1
48	Seager & Slabaugh (2011)	N	N	N	N	M	1
49	Silberberg (2000)	N	M	M	N	S	4
50	Spencer, Bodner & Rickard (1999)	N	N	M	N	N	1
51	Spencer, Bodner & Rickard (2012)	N	N	M	N	S	3
52	Stoker (2010)	N	N	N	N	N	0
53	Stoker (2016)	N	N	N	N	N	0
54	Timberlake (2010)	N	M	N	N	N	1
55	Tro (2008)	S	S	M	N	S	7
56	Umland & Bellama (1999)	N	N	S	S	M	5
57	Whitten, Davis, Peck & Stanley (2010)	N	N	N	N	N	0

(continued)

(continued)

58	Zumdahl & DeCoste (2015)	S	S	M	N	S	7
59	Zumdahl & Zumdahl (2007)	S	S	M	N	N	5
60	Zumdahl & Zumdahl (2014)	S	S	M	N	M	6

[a]Criteria: (for details see text)
1. Objectivity
2. Scientific method
3. Scanning tunneling microscopy (STM)
4. Atomic force microscopy (ATM)
5. From representation to presentation: Scientific progress at a crossroads
S = Satisfactory, M = Mention, N = No mention
[b]Points
S = 2 points, M = 1 point, N = 0 points

Index

© Springer International Publishing AG 2018
M. Niaz, *Evolving Nature of Objectivity in the History of Science and its Implications for Science Education*, Contemporary Trends and Issues in Science Education, DOI 10.1007/978-3-319-67726-2

Printed in the United States
By Bookmasters